AMERICAN HANDGUN PATENTS 1802-1924

by Joseph J · Macewicz

MUSEUM
RESTORATION
SERVICE
1977

Canadian Cataloguing in Publication Data

Macewicz, Joseph John, 1929-
American handgun patents, 1802-1924
Includes index.
ISBN 0-88855-172-X pa.:$4.95

1. Firearms -- Patents. 2. Firearms, American.
I. 0100-B-5675 Museum Restoration Service. II. Title.
TS537 M33 683' .4'00272 C77-000147-5

Printed in Canada by
Museum Restoration Service
Bloomfield, Ontario
K0K 1G0

INTRODUCTION

"any person who invents or discovers any new and useful process, machine, or composition of matter, or any new and useful improvements thereof, may obtain a patent."

With the above terms of reference as the official United States Government guideline for the issuance of a patent, there have been nearly four million patents issued since the patent office first opened its doors.

The early patents, that is, those issued between 1802 and 1837, did not have a patent number assigned to them. It was the patent office practice to use dates only and they are now known as "Name & Date" Patents. If these patents were reissued, a "Reissue Number" was assigned. This may cause some confusion when using this listing, as, if a patent was reissued it will have a lower number than those adjacent to it. But each has been cross-referenced to the original date or patent number.

The key word in the terms of reference is "useful" and an additional statute notes "The term useful in this connection refers to the condition that the subject matter has a useful purpose and also operativeness; that is, a machine which will not operate or perform the intended purpose would not be called useful."

The United States Patent Office maintains 310 classes and 64,000 sub-classes of patents which cover many diversified fields of endeavour. This at times gives rise to confusion. When an invention is patented in one field but has application in another, it makes location of applicable patents most difficult. There is for example a case where a designer of linkages for power transmissions got an idea for using the toggle action employed in a lever-action rifle in a similar manner. This led to a patent being issued for a firearm in the transmission sub-classification. In examining the patent it is recognized that the idea could have just as easily been placed in the sub-class dealing with firearms but it wasn't. Hence, there are a number of areas of Patent Office records which must be considered.

Basic patents on Firearms can be found in the following classes and sub-classes.

CLASS	SUB-CLASS
24	2.5
29	1.1, 1.11, 1.2, 1.21, 1.22, 1.23, 1.31, 1.32
30	164.9
33	233
42	1, 2, 3, 4, 5, 6, 7, 8, 9, 10, 11, 12, 13, 14 15, 16, 17, 18, 19, 20, 21, 22, 23, 24, 25, 26 27, 28, 29, 30, 31, 32, 33, 34, 35, 36, 37, 38 39, 40, 41, 42, 43, 44, 45, 46, 47, 48, 49, 50 51, 52, 53, 54, 55, 56, 57, 58, 59, 60, 61, 62 63, 64, 65, 66, 67, 68, 69, 70, 71, 72, 73, 74 75, 76, 77, 78, 79, 80, 81, 82, 83, 84, 85, 86 87, 88, 89, 90, 91, 92, 93, 94
73	167
81	3.05
89	1, 1.7, 4, 9, 11, 12, 13, 14, 15, 16, 17, 18, 19 20, 21, 22, 23, 24, 25, 26, 27, 28, 29, 30, 31, 32 33, 34, 35, 36, 37, 38, 39, 40, 41, 42, 43, 44, 45, 46 47, 124, 125, 126, 127, 128, 129, 130, 131, 132, 133 134, 135, 136, 137, 138, 139, 140, 141, 142, 143 144, 145, 146, 147, 148, 149, 150, 151, 152, 153, 154 155, 156, 157, 158, 160, 161, 162, 163, 164 165, 166, 167, 168, 169, 170, 171, 172, 173, 174, 175 176, 177, 178, 179, 180, 181, 182, 183, 184, 185, 186 187, 188, 189, 190, 191, 192, 193, 194, 195, 196 197, 198, 199
90	28.1
102	24, 25, 26, 27, 28, 29, 30, 32, 38, 39, 40, 41 42, 43, 44, 45, 46, 47, 48, 52, 63, 65.2, 86.5 89, 92.1, 92.7
124	4, 11, 12, 17, 28, 45
211	64
222	79
224	1, 2, 3, 14, 15, 16, 17, 18, 19, 20, 21, 22, 23 24
240	6.41, 10.62

This study has been limited to handguns and the period which showed the most creativity. Information has come from numerous sources but the author is indebted particularly to the U.S. Government Patent Office and the Milwaukee Public Museum Library.

J.J. Macewicz

AMERICAN HANDGUN PATENTS

July 6, 1802
to
March 17, 1834

July 6, 1802, **EDWARD WEST**, New Liberty, KY. Gunlock.

June 10, 1818, **ARTEMUS WHEELER**, Concord, MA. Repeating gun to discharge 7 or more times.

June 19, 1822, **JOSHUA SHAW**, Philadelphia, PA. Percussion gun improvements.

October 31, 1822, **ALBERT HOLCOMB**, Litchfield, NY. Firearm.

April 12, 1823, **JOSEPH FINCH**, New York, NY. Percussion gun and pistols.

May 20, 1824, **J. COOK**. Gunlock.

June 2, 1825, **JAMES TYRER**, Petersburg, PA. Percussion guns.

November 24, 1826, **FORDYCE RUGGLES**, Hardwicke, MA. Firearm.

February 3, 1827, **SIMON CROMWELL**, Edgecomb, MA. Gunlock.

February 20, 1827, **WILLIAM A. HART**, Fredonia, NY. Percussion gunlock.

July 10, 1827, **MARVEL DAVIS**, Mayville, NY. Percussion gunlock.

September 11, 1827, **JAMES B. LOWRY**, Mayville, NY. Magazine percussion gunlock.

October 16, 1827, **JOHN AMBLER, JR.**, New Berlin, NY. Lever percussion gunlock.

November 10, 1827, **JOSEPH SHATTUCK**, Jefferson County, OH. Percussion gunlock.

May 3, 1828, **OREN MOSES**, Malone, NY. Firearm percussion lock.

May 24, 1828, **JOSEPH LAWRENCE**, New Berlin, NY. Percussion gunlock.

May 29, 1828, **NATHANIEL SALTONSTALL**, New London, CT. Gunlock.

August 29, 1828, **SAMUEL L. FARIES**, Middleton, OH. Waterproof, self-priming, percussion gunlock.

May 7, 1829, **J. SHAW**. Gun and firearm.

May 7, 1829, **HENRY RODGERS**, Middletown, OH. Revolving four barreled gun.

June 11, 1829, No. 203, **JAMES MILLER**, New York City. An improvement in construction of firearms---seven shot cylinder.

October 10, 1829, **SAMUEL L. FAIRES**, Middletown, OH. Eight barreled, percussion gun.

December 7, 1829, **ROBERT EASTMAN**, Brunswick, MA. Firearm.

April 27, 1830, **J. NEWBURY**. Concealed percussion lock.

December 23, 1830, **M. CARLETON**, Percussion lock.

June 13, 1831, **ABEL JUDSON**, New Lebanon, NY. Gunlock.

July 20, 1831, **ALVIN D. CUSHING**, Troy, NY. Cane gun.

August 2, 1831, **E. STRONG**, Claridon, OH. Side percussion lock for rifles.

February 17, 1832, **ISRAEL P. RICHARDSON**, Palmyra, NY. Percussion lock.

February 27, 1832, **RODGER N. LAMBERT**, Upton, MA. Cane gun.

April 9, 1832, **FRANCIS DOWLER**, Wayne, OH. Percussion lock.

October 10, 1832, **PHILIP MORINEAU**, Philadelphia, PA. Firearms, loading and discharging.

October 25, 1832, **DAVID G. COLBURN**, Port Byron, NY. Percussion gunlock.

June 29, 1833, **DAVID G. COLBURN**, Port Byron, NY. Top hammer revolving rifle with two triggers - one to revolve cylinder and one to trip hammer. Instead of ratchets being cut into rear of cylinder, studs project from rear, being engaged by the hand. Reissued as No. 7620X.

August 17, 1833, **ROBERT H. FAUNTLEROY**, Posey County, IN. Gunlock.

March 17, 1834, **JOSHUA SHAW**, Philadelphia, PA. Percussion pistol and whip.

October 7, 1834, **J.B. GRAY**, Fredericksburg, VA. Gunlock.

February 20, 1835, **ROBERT BEALE**, Washington, DC. Percussion lock for guns.

August 17, 1835, **VICTOR M. WALLAC**, West Topsham, VT. Pocket pistol.

February 25, 1836, No. 94,30X, **SAMUEL COLT**, Hartford, CT. Folding trigger, hammer revolver and hammerless rifle with ring-trigger to revolve cylinder and regular trigger to trip hammer.

April 13, 1836, No. 9591X, **B. & B. DARLING**, 6-shot, single action pepperbox with center hammer like old percussion pocket pistols (single-shot type with round butt). Ratchet and pawl like Colt.

April 29, 1837, No. 182, **DANIEL LEAVITT**, Cabotsville, MA. 6-shot, single-action, solid frame revolver with tip-up action released by pressing lever inside front end of trigger guard. Center-hammer. Round-butt.

April 29, 1837, No. 183, **JOHN WEBSTER COCHRAN**, New York City, NY., Revolving hammerless rifle with cylinder revolving horizontally and tip-up barrel. Furnishes a vent in top of barrel above joint with cylinder to allow some of the gases to escape. 9 shots.

April 29, 1837, No. 188, **JOHN W COCHRAN**, New York City, NY.

May 30, 1837, No. 216, **OTIS W. WHITTIER**, Enfield, NH. Ten-shot, hammerless revolving rifle with slotted cylinder surface as means of revolution.

July 5, 1837, No. 254, **GEORGE ELGIN**, New York, NY. Pistol, knife or cutlass.

July 29, 1837, No. 328, **ROBERT W. ANDREWS**, Stafford, CT Pistol sword.

August 31, 1837, No. 364, **SILAS DAY**, New York, NY. Self-loading and self-priming pistol.

November 23, 1837, No. 481, **ROBERT W. LAWTON**, Newport, RI. Pistol saber.

February 15, 1838, No. 603, **EPENSTUS A. BENNETT & FREDERICK P. HERSILAND**, Waterville, ME. An endless chain of steel chambers revolves in a horizontal plane, one by one coming into firing position.

April 5, 1838, No. 677, **HENRY & CHARLES DANIELS**, Chester, CT. Octagonal cylinder revolves in horizontal plane, Under hammer. Revolving rifle.

April 21, 1838, No. 698, **THEDORE F. STRONG**, Northampton, MA. Nine shot, revolving rifle.

April 24, 1838, No. 707, **NICHOLS & CHILDS (RUFUS & EDWARDS)**, Conway, MA. Nine shot, revolving rifle with top hammer of curious construction.

April 25, 1838, No. 713, **MIGHILL NUTTING**, Portland, ME. Hammerless revolving rifle with two triggers.

July 12, 1838, No. 832, **ELIJAH JAQUITH**, Brattleboro, VT. Under hammer, revolving rifle with peculiar ratchet cuts.

March 16, 1839, No. 1,106, **ELON B. BUTTERFIELD**, Brattleboro, VT. Filler block to make-up length differences of cylinders, when changing calibers.

April 25, 1839, No. 1,134, **DAVID EDWARDS**, Zanesville, Morgan County, OH, Six shot single action, side-hammer revolver with round butt.

August 29, 1839, No. 1,304, **SAMUEL COLT**, Paterson, NJ. Revolving rifle with gas check. Open top wedge-locked frame. Also a "Patterson" type revolver and a cap loader. Also a bullet loader, etc.

April 16, 1845, No. 3,998, **ETHAN ALLEN**, Norwich, CT. Double action top hammer, six shot pepperbox with ring-trigger.

September 2, 1847, No. 5,316, **L H GIBBS (LUCIUS)**, Oberlin, OH. Top hammer with mainspring hung in front of hammer and over barrel. Seven shot rifle with revolver cylinder.

October 24, 1848, No. 124, **SAMUEL COLT**, Hartford, CT. Reissue.

May 15, 1849, No. 6,453, **J. POST**, Newark, NJ. Double action, ring trigger, pepperbox with bottom hammer reciprocating in long axis of piece and striking nipple at six o'clock.

August 28, 1849, No. 6,669, **EDWIN WESSON**, Northborough, MA. Cylinder revolves by bevel gears. 3 shot revolver, tip-up, with side hammer.

September 18, 1849, No. 6,723, **GEORGE LEONARD, Jr.** Shrewsbury, MA. Double action, hammerless, pepperbox with revolving hammer-nose placed on a shaft in axis of cylinder. Fixed barrels.

December 4, 1849, No. 6,925, **JACOB PECARE & JOSIAH M. SMITH**, New York, NY. Concealed trigger.

December 18, 1849, No. 6,960, **CHRISTIAN SHARPS**, Washington, DC. Four shot pepperbox with stationary barrels and revolving hammer-nose hung on a shaft in axis of barrels and struck from rear by a hammer of usual type. Single-action.

March 26, 1850, No. 7,218, **HANS IVERSON**, New York, NY. Twelve shot revolver with cylinder revolving in horizontal plane; under hammer; saw handle.

April 23, 1850, No. 7,300, **DEXTER H. CHAMBERLAIN**, Boston, MA. Three shot ring-trigger, pepperbox with straight handle.

July 9, 1850, No. 7,493, **GEORGE LEONARD**, Jr., Shrewsbury, MA. Six shot hammerless, ring trigger pepperbox.

August 13, 1850, No. 172, **JACOB PECARE & JOSIAH M. SMITH**, New York, NY. (Reissue) Folding trigger for double action, hammerless pepperbox.

September 3, 1850, No. 7,613, **SAMUEL COLT**, Hartford, CT. 1850 model Colt with center hammer, square-backed trigger guard and a solid tip-up frame hinged at cylinder barrel junction. Oval cylinder notches.

September 10, 1850, No. 7,629, **SAMUEL COLT**, Hartford, CT. Six shot revolver with square cylinder notches and open top wedge locked frame.

November 26, 1850, No. 7,802, **JOSHUA STEVENS**, Chicopee Falls, MA. Six shot hammerless tip-up revolver hinged in rear of cylinder. Double action. No loading lever.

January 7, 1851, No. 7,887, **STANHOPE W. MARSTON**, New York, NY. New lock action for double action, ring trigger, top hammer, pepperbox.

January 7, 1851, No. 7,894, **JAMES WARNER**, Springfield, MA. Lock action for single action side hammer, solid frame, solid top, round butt, revolver.

July 8, 1851, No. 8,210, **PATRICK W. PORTER**, Memphis, TN. Revolving breechloader.

July 15, 1851, No. 8,229, **JAMES WARNER**, Six shot, side-hammer, single action revolver with solid frame, closed top. No loading lever.

October 7, 1851, No. 8,412, **JOSHUA STEVENS**, Chicopee Falls, MA. Side hammer, tip-up revolver with frame hinged at rear of cylinder. Loading lever. Single action.

June 1, 1852, No. 8,982, **HENRY S. NORTH**, Middletown, CT. & **CHAUNCEY D. SKINNER**, Haddam, CT. Solid frame (closed top) top-hammer, double action revolver. Hammer cocked and cylinder revolved by depressing trigger guard. Trigger trips hammer.

May 3, 1853, No. 9,694, **ROBERT ADAMS**, London, England. Hammer safety to hold it back while arm is charged. Also a cylinder stop as projection on trigger.

May 3, 1853, No. 9,701, **CHARLES N. TYLER**, Worcester, MA. Repeating arm.

August 9, 1853, No. 9,922, **GEORGE LEONARD**, Shrewsbury, MA. Lock mechanism for double-action hammerless pepperbox.

August 9, 1853, No. 9,929, **JOSHUA STEVENS**, Chicopee Falls, Ma. Hammer (side) and hammerless revolver, frame hinged at rear.

November 22, 1853, No. 10,259, **M.L. ROOD**, Marshall, MI. Six shot single action, side hammer revolving rifle.

April 25, 1854, No. 10,812, **JOSIAH ELLS**, Allegheny, PA. Double action open top, solid frame revolver, Six shots. Hammer without cocking spur.

April 25, 1854, No. 10,821, **CHARLES BUSS**, Marlbourough, NH. Trigger guard also "presses index holder against catch wheel." Solid frame, single action revolver with closed top.

May 16, 1854, No. 10,930, **JEREMIAH PECK**, New Haven, CT. (Assd. to self and Charles F. Grilly) Cylinder release hammer safety, and cap protector for solid top, solid frame, single action revolver. Cylinder revolved by lever action on a hand placed at front end. Lever manually operated by thumb before cocking hammer.

June 27, 1854, No. 11,197, **EDWARD LINDNER**, New York, NY. Repeating pistol.

August 1, 1854, No. 11,419, **JOSIAH ELLS**, Allegheny, PA. Double action, hammer revolver. Open top solid frame. Also cylinder lock.

August 1, 1854, No. 11,447, **ELI WHITNEY**, Whitneyville, CT. Ring trigger, five shot, solid frame, solid top, double action revolver.

August 8, 1854, No. 11,470, **ISAAC W. BROWN**, W. Springfield, MA. Double action, top hammer, eight shot solid top, solid frame revolver.

September 19, 1854, No. 11,698, **WILLIAM HY. MORRIS**, New York, NY. Revolver with cylinder swinging to side. First sheet of description missing and cut unintelligible.

September 26, 1854, No. 11,715, **FORDYCE BEALS**, New Haven, CT. Ring trigger, revolver, trigger operating 2 ratchets which engage cuts on the left side of cylinder, not on its rear, Six shot. Solid frame, solid top.

November 7, 1854, No. 11,917, **WENDELL WRIGHT**, New York City, NY. Cylinder revolves in a vertical plane. Revolver, Eight shots.

January 2, 1855, No. 12,189, **JOSHUA STEVENS**, Chicopee Falls, MA. Six shot, double action, solid frame, solid top revolver with (right) side hammer and a single nipple which must be recapped at each shot.

January 16, 1855, No. 12,235, **EDMUND H. GRAHAM**, Bedford, ME. Hammerless revolving rifle. Cylinder revolves in horizontal plane with chambers hanging in a vertical plane. When a chamber approaches the barrel an arm throws it upward into horizontal plane.

January 30, 1855, No. 12,328, **A.O.-P.H. SCHORN**

January 30, 1855, No. 12,328, **A.O.-H.P. SCHORN**, Murphreesborough, TN. Combination cane, dagger, and revolver.

February 27, 1855, No. 12,470, **JOHN HOLLINGSWORTH & RALPH S. MERSHON**, Zanesville, OH. A spiral clockspring behind hammer stores up power which is released by each pull of the trigger to rotate cylinder and trip hammer.

February 27, 1855, No. 12,471, **JOHN HOLLINGSWORTH & RALPH S. MERSHON**, Zanesville, OH. Solid frame, double action, hammerless revolver with solid top. Two lock types shown.

March 20, 1855, No. 12,555, **FREDERICK D. NEWBURY**, Albany, NY. Open top, center hammer (without cocking spur) double action revolver with hand on right side and a ratchet plate separable from rear of cylinder.

April 3, 1855, No. 12,649, **ROLLIN WHITE**, Hartford, CT. Pulling down on bullet seating lever operates a rack and pinion arrangement which automatically cocks hammer and transfers a paper cartridge into chamber from magazine attached to left side of barrel just ahead of cylinder.

June 12, 1855, No. 13,039, **FREDERICK D. NEWBURY**, Albany, NY. Cylinder only shown.

July 31, 1855, No. 126, **FREDERICK D. NEWBURY**, Albany, NY. (reissue) Barrel hinged in front of hammer. Bottom lock not in front of cylinder but about midway of it. (Smith and Wesson tip-up type.) Metallic cartridge.

September 18, 1855, No. 13,581, **WILLIAM W. MARSTON**, New York

City, NY. Three shot double action, top hammer pepperbox.

September 18, 1855, No. 13,582, **FREDERICK NEWBURY**, Albany, NY. Lower catch at front end of cylinder (same location as Smith and Wesson.)

September 18, 1855, No. 13,582, **FREDERICK NEWBURY**, Albany, NY. Cam-wheel turns cylinder, and a new type, barrel catch.

October 9, 1855, No. 13,660, **WILLIAM MOUNT STORM**, New York, NY. A blunderbuss of huge bore revolving cylinder in rear, apparently to furnish priming only. Whale gun. Ring trigger and top hammer.

December 25, 1855, No. 13,999, **E.K. ROOT**, Hartford, CT. (Samuel Colt, witness) Ring trigger, double action revolver with (left) side hammer. Stud on top of tang passes through frame to engage slots on cylinder and rotate same.

January 15, 1856, No. 14,118, **E.T. STARR**, New York City, NY. Six shot double action, top hammer pepperbox with round butt.

March 11, 1856, No. 14,420, **WILLIAM MOUNT STORM**, New York, NY. Solid frame, open top, ring trigger, revolver with hinge at front of trigger below. Cylinder surrounded by a circular housing which furnished anchorage for top lock at front end of cylinder.

March 11, 1856, No. 14,406, **FREDERICK NEWBURY**, Albany, NY. Seven shot, magazine revolver with 2 chambers in cylinder.

March 11, 1856, No. 14,420, **WILLIAM MOUNT STORM**, New York City, NY. Barrel hinged below and in front of cylinder and catch is on top of frame in front of cylinder.

March 25, 1856, No. 14,488, **FREDERICK WILLIAM HOFFMAN & GUSTAV A. BLITTKOWSKI**, New York, NY. Needle fire, revolving rifle with side (right) hammer and a reciprocating action to cylinder which produces tight joint at time of firing.

April 22, 1856, No. 14,710, **FREDERICK WILLIAM HOFFMAN**, New York, NY.

April 29, 1856, No. 14,780, **SIMON F. STANTON**, Manchester, NH. Hammerless revolving rifle with cylinder revolving in horizontal plane, chambers hanging in vertical plane, and being thrown into horizontal plane at proper time.

May 6, 1856, No. 14,820, **GEORGE LEONARD**, Shrewsbury, MA. Five shot ring trigger, double action pepperbox with revolving hammer situated in axis of cylinder, making one-fifth revolution each time trigger is pulled (barrels remain fixed). Chamber section of cylinder tips down for capping (hinged on bottom of frame with latch on top.) Barrel section may be unscrewed from chamber section.

May 20, 1856, No. 14,905, **SAMUEL COLT**, Hartford, CT. Single action, open top, solid frame revolver with revolution effected by a sliding bar attached to top of hammer and having a reciprocating motion in a groove in top strap. Lug on lower side of bar engages helical cuts on cylinder surface.

June 3, 1856, No. 15,032, **F.B.E. BEAUMONT**, England. Double action solid frame, solid top revolver much like Adams.

June 10, 1856, No. 15,110, **ALEXANDER HALL**, New York, NY. Revolving rifle with large cylinder holding 20 rounds. Solid frame with solid top. Cylinder axis well below bore of barrel. Top hammer working in horizontal plane.

June 17, 1856, No. 15,144, **HY. S. NORTH**, Middletown, CT. Double action, top hammer, ring trigger revolver with solid top and solid frame.

June 24, 1856, No. 15,202, **J. WARNER**, Springfield, MA. Solid frame, solid top revolving rifle with a single supple set in frame.

June 24, 1856, No. 15,167, **FORDYCE BEALS**, New Haven, CT. Single action, solid top, solid frame revolver with hand partly exposed on left side of frame.

July 22, 1856, No. 15,388, **C.S. PETTINGILL**, New Haven, CT. Double action, hammerless revolver. Solid top, solid frame.

September 16, 1856, No. 15,734, **E.H. GRAHAM**, Manchester, NH. Five shot underhammer, revolving rifle, trigger guard for a mainspring. Cylinder revolves in horizontal plane and chambers are in normal position at all times.

September 30, 1856, No. 15,797, **JOSEPH ADAMS**, Cleveland, OH. Revolving rifle (whose barrels revolve); 3 barrels of equal bore, or four in two sets of different bores. Double action top hammer.

October 21, 1856, No. 15,925, **ALEXANDER LEMAT**, New Orleans, LA. Superposed barrel revolver with large bore barrel underneath, fired by double-nosed hammer.

January 13, 1857, No. 16,367, **ETHAN ALLEN**, Worcester, MA. Hammerless, open top, pepperbox with trigger guard operating a pinion which engages rack on cylinder pin to lock barrel portion to cylinder pin.

January 13, 1857, No. 16,411, **ALFRED TONKS**, Boston, MA. Double action, ring trigger, solid frame, solid top revolver.

January 27, 1857, No. 16,477, **HEINRICH GENHART**, Belgium. Hammerless rifle with cylinder revolving in horizontal plane. Ten shots.

February 10, 1857, No. 16,575, **FRANK S. BRETTELL & JOSEPH B. FRISBIE**, Allegheny, PA. Double action, solid frame, solid top, revolver. Handle like Smith and Wesson tip-up. Top hammer with finger hook so that it may be cocked for single action use.

February 24, 1857, No. 16,683, **SAMUEL COLT**, Hartford, CT. Cylinder stop fitted into base of hammer and acting in slots and notches on rear of cylinder.

March 3, 1857, No. 16,716, **SAMUEL COLT**, Hartford, CT. The typical Colt revolving rifle. Side hammer on left side.

April 14, 1857, No. 17,032, **JOSIAH ELLS**, Pittsburgh, PA. Double action, solid frame, open top revolver like Colt. May be used single action or double action.

April 14, 1857, No. 17,044, **JAMES KERR**, London, England. Device for ramming cartridges into chambers.

April 28, 1857, No. 17,143, **JOSIAH ELLS**, Pittsburgh, PA. Double action, top hammer, solid top, solid frame revolver with handle like old Smith and Wesson.

May 26, 1857, No. 17,359, **FORDYCE BEALS**, New Haven, CT. Single action, solid frame, solid top revolver with pawl partly exposed on left side of frame and top of frame bowed up when hammer pierces it.

May 26, 1857, No. 17,382, **EDWARD LINDNER**, New York City, Improvement on No. 11,197 cocking arm moved from bottom to the top of the action.

May 26, 1857, No. 17,386, **WILLIAM W. MARSTON**, New York, NY. Oscillating cylinder to which is attached a firing knob successively opposite 3 nipples placed on as many barrels in a vertical line.

June 30, 1857, No. 17,698, **JACOB SHAW**, Jr. Hinckley, OH. Compact, double action, hammerless revolver with hammer striking chamber at six-o'clock and barrel at bottom of frame instead of top.

July 28, 1857, No. 17,904, **JAMES WARNER**, Springfield, MA. Special hand device for "Warners Pat" single action, side hammer, solid frame, solid top revolver.

August 4, 1857, No. 17,915, **IRA BUCKMAN**, New York, NY. Cane gun.

October 20, 1857, No. 18,486, **G.R. CROCKER**. Safety device for Colt single action, open top revolver.

November 24, 1857, No. 18,678, **SAMUEL COLT**, Hartford, CT. Solid frame, open top revolver with cylinder pin inserted from rear. Six shot.

December 15, 1857, No. 18,836, **ETHAN ALLEN**, Worcester, MA. Solid top, solid frame revolver with improvements in turning and locking devices.

February 9, 1858, No. 19,327, **FREDERICK D. NEWBURY**, Albany, NY. Double action, solid top, solid frame revolver. Hammer without cocking spur. Six shots. Rammer works by rack and pinion.

February 9, 1858, No. 19,328, **JOHN F. THOMAS**, Ilion, NY. Cane gun.

March 23, 1858, No. 19,674, **ROBERT BECKWITH**, New York, NY. Cane guns.

March 23, 1858, No. 19,739, **FREDERICK D. NEWBURY**, Albany, NY. Break open, side hammer revolver. Latch in front of cylinder and hinge below and in front.

April 6, 1858, No. 19,868, **H.S. NORTH**, Middleton, CT. Rack and pinion arrangement for ramming cartridge into chamber.

April 13, 1858, No. 19,961, **ROLLIN WHITE**, Hartford, CT. A five shot, double action, top hammer revolver for teat cartridges with chambers enlarged at rear for purpose of allowing explosion to drive cylinder forward and make gas tight joint with barrel. Resembles Allen & Thurber double action, top-hammer revolver.

May 5, 1858, No. 20,144, **SAMUEL COLT**, Hartford, CT. A typical side-hammer, pocket model revolver.

May 4, 1858, No. 20,160, **BENJAMIN JOSLYN**, Stonington, CT.

June 8, 1858, No. 20,496, **MOSES KINSEY**, Newark, NJ. Single action, solid top, solid frame revolver.

June 15, 1858, No. 20,607, **F.H. HARRINGTON**, MA. (Springfield). Looks like original model Smith and Wesson tip-up revolver with hinged hammer, nose and spring hookcatch. Square butt. Ass'd. S & W.

June 29, 1858, No. 20,765, **FREDERICK D. NEWBURY**, Albany, NY. Double action, solid frame, solid top revolver.

July 27, 1858, No. 21,054, **EDWARD A. RAYMOND & CHARLES ROBITAILLE**, Brooklyn, NY. Double action, hammerless, solid frame, solid top revolver. Six shots.

August 17, 1858, No. 21,188, **WILLIAM H. ELLIOT**, Plattsburg, NY. The four shot, ring trigger pepperbox of later days.

August 17, 1858, No. 21,215, **JOSEPH RIDER**, Newark, OH. Single action, solid top, solid frame, center hammer revolver. Grip looks like that of Elliot pepperbox.

September 7, 1858, No. 21,400, **ETHAN ALLEN**, Worcester, MA. Cylinder pin made square just below cylinder barrel joint so as to deflect gases and keep them from fouling pin and making its rotation difficult. Gurards on barrel in front of cylinder to protect hand of shooter in case of discharge of a chamber not in line with barrel.

September 14, 1858, No. 21,478, **FORDYCE BEALS**, New Haven, CT. Pocket revolver like Colt side hammer type, but with center hammer.

September 28, 1858, No. 21,623, **WILLIAM PALMER** New York, NY. Six shot double action, center hammer, (without cocking spur), revolver. Solid frame, solid top. Finger hold on trigger guard.

September 28, 1858, No. 204, **FREDERICK D. NEWBURY**, Albany, NY. (Reissue.) Closely like No. 20,765.

October 12, 1858, No. 21,730, **THOMAS K. AUSTIN**, New York, NY. Six shot revolving rifle. Double action, hammerless, with reciprocating hammer traveling in long axis of arm.

November 9, 1858, No. 22,005, **ETHAN ALLEN**, Worcester, MA. Typical Allen & Wheelock type single action, solid frame, solid top, revolver with cylinder pin inserted from rear. Apparently for rim-fire cartridges. (Right) side hammer.

December 14, 1858, No. 633, **ETHAN ALLEN**, Worcester, MA. (Reissue.) No. 3,998.

December 21, 1858, No. 22,348, **E. CLAUDE**, New York, NY. Six shot, double action revolver for rim fire cartridges with top latch permitting barrel and cylinder to swing to left on the arc of a circle with axis just above trigger where a catch on right side of frame is released. Projecting cylinder stops like Lefaucheux. Apparently no ejection arangements.

December 28, 1858, No. 22,412, **JOHN WEBSTER COCHRAN**, New York, NY. Six shot hammerless percussion revolver. Cylinder rotated by bevel gearing cut radially around cylinder pin engaged by another gear rotating in long axis of arm and actuated by trigger. No hammer mechanism made out. Hinge at top rear indicates tip-up action.

January 4, 1859, No. 22,511, **C.S. PETTINGILL**, New Haven, CT. Double action, hammerless revolver with solid top, solid frame, separate firing pin and spiral mainspring.

January 18, 1859, No. 22,626, **SAMUEL COLT**, Hartford, CT. Shoulder stock for revolvers.

January 18, 1859, No. 22,627, **SAMUEL COLT**, Hartford, Ct. Canteen, shoulder stock for revolver.

January 18, 1859, No. 22,666, **HENRY S. NORTH**, **EDWARD SAVAGE**, Middletown, CT. Single action top hammer, revolving rifle.

January 25, 1859, No. 22,753, **CHRISTIAN SHARPS**, Washington, DC. Breechloading Firearms.

February 1, 1859, No. 652, **JOSIAH ELLS**, Allegheny, PA. Reissue of 10,812.

February 8, 1859, No. 22,905, **JOHN WALCH**, New York, NY. Revolver with long chambers holding two cartridges each. Two nipples on each chamber. Two hammers and one trigger.

March 1, 1859, No. 23,087, **W.C. HAYNES**, Melrose, TX. Solid frame, open top, revolving rifle with a multi-bored barrel and a large cylinder containing four or five chambers coming in turn opposite the five bores of the barrel.

April 19, 1859, No. 23,711, **JACOB RUPERTUS**, Philadelphia, PA. Six shot, solid frame, center hammer revolver. Cap magazines in stock automatically feeds caps to hammer when same is cocked.

May 3, 1859, No. 23,861, **JOSEPH RIDER**, Newark, OH. The typical pocket type, double action "Remington Rider" revolver save that frame is bowed where hammer pierces it.

June 7, 1859, No 24,274, **THOMAS BAILEY**, New Orleans, LA. Break open revolver with top latch and bottom hinge located just as in later cartridge types. Double action. Side hammer.

June 28, 1859, No. 24,573, **HENRY S. NORTH** and **JOHN O. COUCH**, Middletown, CT. Hammerless revolving "trap" pistol for game shooting.

July 5, 1859, No. 24,666, **HORACE SMITH** and **DANIEL B. WESSON**, Springfield, MA. Spring hook-catch remains but hammer now has solid nose. Square butt.

July 12, 1859, No. 24,726, **WILLARD C. ELLIS** and **JOHN N. WHITE**, Springfield, MA. Cylinder only shown.

July 26, 1859, No. 783, **STANHOPE W. MARSTON**, New York, NY. (Reissue.) Reissue of No. 7,887.

August 2, 1859, No. 24,936, **ALBERT V. HILL**, Limestone, NY.

August 2, 1859, No. 24,942, **S.C. LEWIS** and **F.P. PFLEGHAR**, Whitneyville, CT. Solid frame, single action, sheath trigger revolver resembling Colt side-hammer type, but with center hammer.

September 6, 1859, No. 806, **JOSIAH ELLS**, Allegheny, PA. (Reissue.) Reissue of No. 11,419 (?)

November 1, 1859, No. 846, **E.K. ROOT**, Hartford, CT. Reissue. Ring-trigger actuates stud which engages through frame into helical grooves on cylinder face and revolves same. Side hammer on left side.

December 27, 1859, No. 26,641, **JOSEPH GRULER** and **A. REBETEY**, Norwich, CT. The cut shows the typical single-action center hammer Bacon revolver with left side plate, center hammer, square butt and guarded trigger.

No. 25,259, **JOSEPH GRULER** and **A. REBETEY**, Norwich, CT.

January 24, 1860, No. 26,919, **WILLIAM H. MORRIS** and **CHARLES L. BROWN**, New York, NY. A fixed cylinder with six chambers connected to a single barrel by a cone with passages to conduct charges from chambers drilled at an angle and converging on the single barrel. Prevents escape of gas at barrel cylinder joint and premature discharges. (Lyman's patent 1857 somewhat similar.) Rifle form.

February 21, 1860, No. 265, **JOSIAH ELLS**, Allegheny, PA. (Reissue) Reissue of No. 17,032.

March 20, 1860, No. 27,518, **WILLIAM H. BELL**, Baltimore, MD. Solid frame revolver with barrel secured to frame by catches in front of cylinder, above and below. Hollow cylinder pin contains a supply of caps which are automatically put into place each time hammer is cocked. Single action. Center hammer.

March 20, 1860, No. 27,526, **J. MASLIN COOPER**, Pittsburgh, PA. Double action, solid top, solid frame, six shot, revolver (used single action or double action(. May also be used with a cylinder adapted to rim-fire cartridges.

April 10, 1860, No. 27,868, **F.D. NEWBURY**, Albany, NY. Double action, solid top, solid frame revolver with butt like Colt. May be used single action or double action. Has only a half trigger guard which makes a circle with trigger as rear half when trigger is in forward position.

May 1, 1860, No. 278, **F.D. NEWBURY** Albany, NY. Reissue. Shows cylinder only, with safety shield covering nipples and a gate, the opening of which exposes one of these.

May 15, 1860, No. 28,331, **EDWARD SAVAGE** and **H.S. NORTH**, Middletown, CT. Civil War type "Savage" revolver. Ring trigger operates cylinder. Side hammer. Single action.

May 22, 1860, No. 28,437, **ABRAM J. GIBSON**, Shrewsbury, MA. (A.S. Spencer, witness). Double action, hammerless revolver with barrel which is hinged in front of cylinder about in line with its center. Cylinder contains three chambers and revolves on a pin located about where bottom chamber would ordinarily lie. (Makes part revolution only, then returns to original position.)

May 29, 1860, No. 28,460, **W.H. ELLIOT**, Plattsburg, NY. Four barrel pepperbox with sliding catch forward of trigger which released barrels so they may tip down, around an axis above trigger.

May 29, 1860, No. 28,461, **WILLIAM H. ELLIOT**, Plattsburg, NY. Four shot, ring trigger, pepperbox with barrels slotted exteriorly and rotated by cam action. Double action.

July 3, 1860, No. 28,951, **ETHAN ALLEN**, Worcester, MA. Breech plate made with projection in form of inclined plane, causing cylinder to revolve freely at first movement of hammer. (Allen and Wheelock type of single action rim fire revolver.)

July 10, 1860, No. 29,126, **ABRAM J. GIBSON**, Shrewsbury, MA. (R.S. Spencer, witness.) Metallic cartridge, single action, rim fire revolver with lock located as in Smith and Wesson tip-up but hinged just in front of hammer.

July 17, 1860, No. 29,213, **CHARLES R. ALSOP**, Middletown, CT. Closely resembles the Savage revolver, but cylinder revolves on cocking; hammer and a cam pushes cylinder against barrel to form gas-tight joint at same time.

August 7, 1860, No. 29,538, **CHARLES R. ALSOP**, Middletown, CT. Improved rammer.

August 21, 1860, No. 1,029, **S.W. MARSTON**, Stanhope, NY. (Reissue) Reissue of No. 783.

August 21, 1860, No. 29,676, **ARMENIUS DAVIS**, Shelbyville, IN. Cane guns.

September 4, 1860, No. 29,864, **J. MASLIN COOPER**, Pittsburgh, PA. Open top, solid frame, single action revolver, wedge fastened and closely resembling Colt.

September 18, 1860, No. 30,079, **DANIEL MOORE**, Brooklyn, NY. Cylinder swings to right far enough to permit loading of one chamber at a time. Metallic cartridge, single action, open top, rim fire, with center hammer and square butt.

October 2, 1860, No. 30,245, **E.A. PRESCOTT**, Worcester, MA. Specifications mixed with some other invention. Single action, rim fire, solid frame, solid top, revolver with square butt.

October 2, 1860, No. 30,260, **AUGUST SPELLIER**, Philadelphia, PA. Eight shot revolving rifle using two cylinders, one for loose ammunition and one for metallic cartridges. Single action, rim fire, solid frame, solid top.

October 9, 1860, No. 30,309, **ABRAM J. GIBSON**, Shrewsbury, MA. Cylinder swings to right a short distance, like Moore's, permitting single loading. Metallic cartridge, single action, rim fire, solid frame, solid top with square butt and center hammer. Sheath trigger.

October 9, 1860, No. 1,059, **HORACE SMITH** and **DAVID B. WESSON**, Springfield, MA. (Reissue) First model catch, a long, flat spring extending along lower frame and hooking over barrel lug.

October 23, 1860, No. 30,494, **F.D. NEWBURY**, Albany, NY. Six shot, solid frame, open top, revolver. Double action, center hammer. Trigger as in No. 27,868. Numerous new lock devices.

November 6, 1860, No. 30,602, **JOHN ADAMS**, Dolston, England. (Assigned to Thomas Poultney, Baltimore, Md.) Typical "Adams Patent" revolver. Double action, center hammer, Solid top, solid frame.

November 27, 1860, No. 30,765, **CHRISTIAN SHARPS**, Philadelphia, PA. (Harry Howson, attorney.) Cylinder with perforated base cap which screws on, hammer passing through perforations.

December 4, 1860, No. 30,843, **EBEN T. STARR**, New York, NY. Starr's break open, double action, army revolver.

December 18, 1860, No. 30,990, **SMITH AND WESSON**, Springfield, MA. Later type latch catch. Breech plate behind cylinder.

April 6, 1861, No. 32,003, **EDWARD B. SAVAGE**, Cromwell, CT. Shoulder stock for Savage, navy revolver.

May 14, 1861, No. 1,329, **CHARLES R. ALSOP**, Middletown, CT. Reissue.

May 14, 1861, No. 32,333, **CHARLES R. ALSOP**, Middletown, CT. Like 29,213 with ring trigger added, same operating a cam which wedges cylinder against barrel at time of discharge.

July 2, 1861, No. 32,685, **J.A. DEBRAEME**, New York, NY. Barrels show two sets of longtitudinal ventilating slits, encircling it (about 10 narrow slits per set), and beginning about one inch ahead of cylinder, with space of about one-half inch between the two, and ending about one-half inch short of muzzle. Extra trigger in front of trigger guard holds cylinder locked until grasped for firing. Single action, solid top, solid frame. Square butt.

September 14, 1861, No, 2,324, **ETHAN ALLEN**, Worcester, MA. Reissue.

September 14, 1861, No. 33,328, **ETHAN ALLEN**, Worcester, MA. (Sullivan Forehand, witness.) Special cylinder pin and rammer release. Rammer works by rack and pinion. Single action, Solid top, solid frame with center hammer, sheath trigger and square butt like later metallic models.

October 1, 1861, No. 2,378, **W.H. ELLIOT**, Plattsburg, NY. Reissue.

October 1, 1861, No. 33,382, **W.H. ELLIOT**, Plattsburg, NY. (S. Remington, witness.) Improvements on No. 28,461. Claim witnessed by I. Harrington.

October 22, 1861, No. 33,509, **ETHAN ALLEN**, Worcester, Ma. (Hy. C. Wadsworth, witness.) Improved type cylinder stop for the Allen and Wheelock side hammer, rim fire, single action, solid frame revolver.

November 26, 1861, No. 2,766, **CHARLES H. ALSOP**, Middletown, CT. Reissue.

November 26, 1861, No. 33,770, **CHARLES H. ALSOP**, Middletown, CT. Chambers cut with shoulders to carry cylinder against barrel at time of discharge.

December 3, 1861, No. 33,836, **HENRY GROSS**, Tiffin, OH. Single action, rim fire, tip-up revolver like Smith & Wesson. Hammer withdraws cylinder from barrel on cocking, and pushes it forward to make gas-tight joint, in falling.

December 17, 1861, No. 2,928, **WILLIAM H. ELLIOT**, Plattsburg, NY. Reissue.

December 17, 1861, No. 33,932, **WILLIAM H. ELLIOT**, Plattsburg, NY. The old, single action, sheath trigger Remington with square butt and center hammer.

December 24, 1861, No. 3,012, **ANTHONY SMITH**, Hartford, CT. Reissue.

December 24, 1861, No. 34,016, **ANTHONY SMITH**, Hartford, CT. Six shot, single action teat-cartridge revolver. Sheath trigger. Handle resembles old Allen and Wheelock single action type.

December 24, 1861, No. 3,028, **THOMAS S. SHAW**, Philadelphia, PA. Reissue.

December 24, 1861, No. 34,032, **THOMAS S. SHAW**, Philadelphia, PA. (Ass'd. to self and B. Hart.) Top hammer, double action, revolver with round butt just like the Allen & Wheelock type.

January 7, 1862, No. 34,067, **DANIEL MOORE**, New York, NY. Six shot, single action revolver with trigger guard. Cylinder slots midway of cylinder which is fluted, save for a collar around middle containing the slots. Chambers rifled at forward end for two-fifths of their length.

January 7, 1862, No. 34,093, **WILLIAM JONES PITT**, Middletown, CT. Revolver with cylinder pin having pinion attached and operated by rack attached to trigger (ring). Apparently hammer is tripped by another trigger not shown. Stock like the "Savage".

January 14, 1862, No. 34,216, **FREEMAN BRADY**, Jr. and **JOHN C. NOBLE**, Washington, PA. Cylindrical magazine in stock feeds cartridges into a chamber on left side of cylinder by spring pressure. Single action type.

January 21, 1862, No. 34,226, **CHARLES R. ALSOP**, Middletown, CT. Five shot, side hammer (top firing), single action, revolver with handle like Savage. Solid frame, solid top.

February 4, 1862, No. 1,268, **ETHAN ALLEN**, Worcester, MA., (Reissue) Approximately same as No. 28,951.

March 11, 1862, No. 34,625, **D.A. COURTER**, Beloit, WI. Pistol and Sword combined.

March 18, 1862, No. 34,703, **CHARLES EDWARD SNEIDER**, Baltimore, MD. Barrel latch and hinge reversed as regards Smith and Wesson practice. Double cylinder, the two sections facing opposite ways. Long nosed hammer, reaching to rear of forward section. When forward section is exhausted, cylinder is slipped out and reversed on cylinder pin, bringing loaded section into play.

March 25, 1862, No. 34,740, **ROBERT J. COLVIN**, Lancaster, PA. Combined sword and pistol.

March 25, 1862, No. 34,803, **CHARLES R. ALSOP**, Middletown, CT. Assigned to Joseph W. Alsop. Improvement on No. 32,333.

April 8, 1862, No. 34,922, **CHARLES DRAEGER**, Indianapolis, IN. Repeating firearm assigned to self and John Ott.

April 22, 1862, No. 35,052, **JAMES A. WHALEN**, Brooklyn, NY. Method of interchanging cylinders quickly.

April 29, 1862, No. 35,067, **ETHAN ALLEN**, Worcester, MA. Revolving rifle with trigger pull adjusted by turning a screw running through hammer, thus varying size of hammer notch. Trigger guard actuates rack and pinion which carry cylinder pin forward under barrel and permit removal of cylinder. Loading gate. Single action.

April 29, 1862, No. 35,093, **LUCIAN GABEL**, Richman, IN. Sword and pistol combined.

May 27, 1862, No. 35,404, **AARON C. VAUGHAN**, Bedford, PA. Double barrel revolver, one barrel set below and to the side of the other. Two hammers and one trigger. Two concentric sets of chambers in same cylinder. Open top, wedge locked, single action, round trigger guard, and square butt.

May 27, 1862, No. 35,419, **CHARLES W. HOPKINS**, Norwich, CT. Like 30,309 in most respects. (Cylinder swings to right part way only.)

June 17, 1862, No. 35,623, **LUCIUS W. POND**, Worcester, MA. Hinge in front of hammer. Latch as in Smith and Wesson.

June 17, 1862, No. 35,657, **JOHN H. VICKERS**, Worcester, MA. Hinge below frame at front. Catch on top frame at rear.

July 1, 1862, No. 35,760, **JAMES JENKINSON**, Brooklyn, NY. Sliding bayonet.

July 29, 1862, No. 35,996, **JAMES B. DOOLITTLE**, Seymour, CT. Magazine firearm.

July 29, 1862, No. 35,999, **GEORGE W.B. GEDNEY**, New York, NY. Solid frame, solid top, single action, round trigger guard. Square butt. Cylinder released by removing cylinder pin.

September 2, 1862, No. 36,358, **JAMES NICHOLAS**, Limestone, NY. Magazine firearm.

September 23, 1862, No. 36,505, **CHRISTOPHER C. BRAND**, Norwich, CT. (E.S. Renwic, attorney.) Revolving rifle with barrel moving horizontally forward through a link action actuated by downward motion of lever (like Winchester lever). Cartridge projects through cylinder. Details of barrel chamber not shown. Six shot, rim fire. Cylinder evidently moved forward by lever to seat front end of cartridge in barrel.

November 4, 1862, No. 36,821, **HENRY S. ROGERS**, Willowvale, NY. Double action, hammerless, solid top, solid frame, revolver. Handle with long curve like Savage.

November 18, 1862, No. 36,984, **STEPHEN W. WOOD**, Cornwall, NY. Single action revolver with trigger guard resembling Lefaucheux type. Hinged loading gate at right. "Open top" type. Also in rifle form. The fixed breech in form of inclined plane causing fired shells to drop back through the loading gate when this is opened.

November 25, 1862, No. 37,004, **THOMAS J. MAYALL**, Roxbury,

MA. Six shot, single action revolver. Cylinder made in two parts, a permanent front section and demountable rear section. Two rear sections furnished, one loose, and one for fixed ammunition. Hammer fitted with adjustable-nose to serve both types.

December 2, 1862, No. 37,059, **JACOB RUPERTUS**, Philadelphia, PA. Hinge just below barrel in front. Latch hooks over rear end of cylinder pin. Forerunner of present break-open types.

December 2, 1862, No. 37,075, **JAMES JENKINSON**, Brooklyn, NY. Single action revolver resembling "Moore" type with trigger guard. A slide moving transversely to left across the breech uncovers a chamber for loading (at right), and prevents cocking of hammer until it is pushed back into place.

December 9, 1862, No. 37,091, **AUSTIN T. FREEMAN**, Binghamton, NY. Remington type, solid frame, solid top, revolver, but double action. Peculiar type main spring.

January 6, 1863, No. 37,329, **FORDYCE BEALS**, New Haven, CT. Improved rammer. Rammer connection for revolving firearm.

January 27, 1863, No. 37,551, **FRANK P. SLOCUM**, Brooklyn, NY. (M.S. Partridge, witness.) Five shots, Hinge at top rear. Lock at bottom front.

February 17, 1863, No. 37,693, **JOHN C. HOWE**, Worcester, MA. Hinge at bottom, front. Latch at top rear. Forerunner of present break-open types.

March 17, 1863, No. 37,921, **SAMUEL REMINGTON**, Ilion, NY. (William H. Elliot, witness.) Typical large single action Remington with solid top, solid frame, center hammer, and square butt. Three groove barrel.

March 24, 1863, No. 37,961, **ALBERT HALL**, Danville, IA. Solid top, solid frame, single action, sheath trigger revolver like old tip-up Smith & Wesson. Teat cartridges loaded from front of cylinder.

April 14, 1863, No. 38,204, **FRANK P. SLOCUM** (A.P. on cut) Brooklyn, NY. Sliding chamber covers for insertion of rim fire cartridges and a fixed piston to act as ejector. Six shot single action, rim fire revolver with trigger guard.

April 28, 1863, No. 38,279, **C.C. BRAND (Christopher)** Norwich, CT. Five shot rim fire revolving rifle. Cylinder shorter than cartridge which projects through it, front end being seated into barrel by motion of a lever located on tang behind hammer, which is thrown upwards and back.

April 28, 1863, No. 38,336, **JAMES REID**, New York, NY. Six shot, single action sheath trigger, solid frame revolver with stock like old Smith and Wesson tip-up. Hinged side plate at right swings up and back in long axis of arm, apparently uncovering a loading gate and also extracting a shell.

April 28, 1863, No. 38,321, **DANIEL MOORE**, Brooklyn, NY. Open top, wedge fastened frame like Colt. Small, round butt. Teat cartridges.

April 23, 1863, No. 38,336, **JAMES REID**, New York, NY. Solid frame, solid top, single action, sheath trigger, like old tip-up Smith and Wesson. Each chamber has a nipple screwed into its breech, which can be removed (unscrewed), and metallic cartridges inserted instead.

June 16, 1863, No. 38,921, **H. SMITH** and **D.B. WESSON**, Springfield, MA. Sliding-breech block moves forward with hammer to support rear of shell at time of firing.

June 16, 1863, No. 38,984, **LUCIUS W. POND**, Worcester, MA. Cut shows cylinder only, apparently loaded at front with teat cartridges.

June 30, 1863, No. 39,032, **JAMES C. CAMPBELL**, New York, NY. Combined pike and revolving firearm.

July 21, 1863, No. 39,318, **WILLARD C. ELLIS** and **JOHN N. WHITE**, Springfield, MA. (Ass'd. to Hy. Reynolds.) Solid frame, tip-up, single action, sheath trigger, somewhat like Smith and Wesson. Cartridges have bases with hemispherical depressions and are loaded from front of cylinder. Hammer tip strikes concave portion of cartridge near where it joins cylinder portion.

August 4, 1863, No. 39,405, **BENJAMIN F. JOSLYN**, Stonington, CT. (Henry Howson, attorney.) Six shot single action, rim fire revolver with ejector pivoted below forward end of barrel and in frame at forward end of cylinder pin, held in place by spring (self-contained). By compressing spring and releasing rear end, this end becomes free to push through chamber on right of cylinder and eject shell. A release allows cylinder to swing out a little way to make this possible.

August 4, 1863, No. 39,406, **BENJAMIN F. JOSLYN**, Stonington, CT. (Henry Howson, attorney.) Six shot, single action rim fire revolver with trigger guard. Made in two sections frame carrying cylinder and stock, carrying hammer and moving parts. Stock like Schofield, Smith and Wesson. Cylinder pin used as ejector by removing it and passing it through a fixed arm attached to barrel (somewhat like Lafaucheux).

August 4, 1863, No. 39,409, **JAMES KERR**, England. This is the Kerr side hammer, double action revolver.

August 18, 1863, No. 39,619, **STEPHEN W. WOOD**, Cornwall, ny. (J.G. Cooper, witness) Solid frame, open top, center hammer, rim fire revolver. Double action with grip like Lefaucheux.

August 25, 1863, No. 39,645, **HENRY GROSS**, Tiffin, OH. Hinge at top front. Catch at bottom front.

August 25, 1863, No. 1,528, **WILLARD C. ELLIS**, Plattsburg, NY. (Reissue, J.N. White.) Cylinder only, for front loading teat cartridges.

August 25, 1863, No. 39,624, **MICHAEL F. GERAGHTY**, Jersey City, NJ. Single action, rim fire revolver, cartridges being loaded from in front. Cylinder is in two parts, cylinder proper and a rear locking ring which is rotated one way to insert cartridge, and opposite way to grab into the circular ring just above the base. (Cartridges rimless type with base smaller than neck, to permit easy loading). Shells are ejected in front using pin carried under barrel for this purpose.

September 1, 1863, No. 39,771, **CHARLES W. HARRIS**, Pittsburgh, PA. Solid frame, open top, wedge lock, double action revolver like Colt in shape. Various lock improvements, also device to put double action out of business when desired.

September 8, 1863, No. 39,825, **RALPH S. MERSHON**, Philadelphia, PA and **JOHN HOLLINGSWORTH**, Zancsville, OH. Coiled power spring in frame below hammer which will cock and fire a loose ammo or metallic cartridg revolver without human effort; also regular mainspring which can be used at will when desired. Open top, solid frame, single action revolver, wedge locked, like Colt.

September 8, 1863, No. 39,850, **H.D. WARD**, Pittsfield, MA. Solid frame, open top, single action, rim fire revolver shaped like Moore's two barrels parallel, two hammers and one trigger so arranged that two shots may be fired simultaneously, or single firing carried out. Eight shots. Wedge lock.

September 8, 1863, No. 39,869, **JOHN H. VICKERS**, Worcester, MA. Basic patent covering Pond, sliding sleeve, revolver.

September 22, 1863, No. 40,021, **J. MASLIN COOPER**, Pittsburgh, PA. No. 29,864 improved.

October 27, 1863, No. 1,558, **ROLLIN WHITE**, Hartford, CT. (Reissue.) Same as No. 1,557.

November 10, 1863, No. 40,553, **JOHN W. COCHRAN**, New York, NY. Single action, solid frame revolver for cartridges of Burnside type. Two rammers on left side, one in front of cylinder and one behind, alternately force in the cartridge and expel the shell, when manually operated. (Work together on rack and pinion.)

November 24, 1863, No. 40,687, **EDMUND H. GRAHAM**, Manchester, NH. Six shot revolver with chambers cut out and showing large part of each cartridge. Action not readily understood from the cut.

January 5, 1864, No. 41,117, **H.A. BRIGGS & SAMUEL S. HOPKINS**, Norwich, CT. (Assd. to selves and C.A. Converse, Norwich, CT.). Single action, rimfire, sheath trigger, metallic cartridge revolver with square butt. Cylinder wings part way to right on a frame.

January 5, 1864, No. 41,184, **DAVID WILLIAMSON**, Brooklyn, NY. Front loading cylinder for teat cartridges.

February 9, 1864, No. 41,472, **JOHN BIRKENHEAD**, Ilion, NY. Hinged trigger cover to swing back over sheath trigger at half cock and make arm safe.

March 1, 1864, No. 41,803, **STEVEN W. WOOD**, Cornwall, NY. Solid frame, single action revolver with front loaded cylinder for cartridges contained in cylindrical tubes.

March 8, 1864, No. 41,848, **BEN KITTREDGE**, Cincinnati, OH. A shield to attach to an open top Colt type, loose ammo revolver between cylinder and fixed breech, to deflect laterally any flare from cap.

March 8, 1864, No. 41,857, **WILLIAM PALMER**, New York, NY. Drawing lacking detail shows a hammerless revolver without trigger operated by a revolving crank handle on right side of frame, cartridges being fed into cylinder through an Accles type feed at about 8 o'clock.

April 19, 1864, No. 42,379, **BENJAMIN F. JOSLYN**, Stonington, CT. (Henry Howson, attorney.) Cylinder made in two positions, rear fixed and actuated by hand, and front moving with barrel horizontally forward through motion of rack and pinion, controlled by downward and forward motion of trigger guard. Extraction accomplished by moving barrel forward.

April 19, 1864, No. 42,435, **THOMAS GIBSON**, Yonkers, NY. Cylinder stop for Starr Army revolver (12 notches).

May 10, 1864, No. 42,648, **WILLIAM H. ELLIOT**, Plattsburg, NY. Four shot pepperbox with hammer visible and regular (one half ring) trigger. Barrel hinge in front of trigger and barrels tip down.

May 10, 1864, No. 42,649, **WILLIAM H. ELLIOT**, Plattsburg, NY. Two vertically reciprocating dogs act as firing pins to discharge two sets of superposed barrels (four shot pepperbox). Two types shown.

May 10, 1864, No. 42,688, **HENRY REYNOLDS**, Springfield, MA. Single action, solid frame, solid top revolver for front loading, mettalic cartridges to be expelled from rear. Typical of a make sold under another name. Sheath trigger and square butt.

May 10, 1864, No. 42,698, **EBEN T. STARR**, New York, NY. Four shot, hammer pepperbox. Hinge below barrels about midway of them. Tips down.

May 17, 1864, No. 42,823, **DAVID WILLIAMSON**, Brooklyn, NY. (Assd. to Moore's Patent Fire Arms Company). Cylinder stop for single action, wedge type, solid frame open top revolver with round butt.

June 7, 1864, No. 1,693, **DANIEL MOORE**, Brooklyn, NY. (Reissue.) Duplicates No. 38,321, five grooves, six shots.

June 7, 1864, No. 1,694, **DANIEL MOORE**, Brooklyn, NY. (Reissue.) Duplicates No. 38,321.

July 5, 1864, No. 1,715, **WILLIAM H. ELLIOT**, Plattsburg, NY. (Reissue.) Duplicates No. 33,932.

July 12, 1864, No. 43,529, **R.D.O. SMITH**, Washington, DC. Six shot single action, rim fire, solid frame solid top sheath trigger, square butt, revolver. Hammer has lips at top and bottom which grip shell and withdraw it after explosion when hammer is cocked. (Both act as firing pins.) Rotating plate with apertures for lips may be thrown in front of hammer if extraction is not desired.

July 19, 1864, No. 43,606, **JACOB RUPERTUS**, Philadelphia, PA. (Harry Howson, attorney.) Eight shot, pepperbox with loading gate. Single action, sheath trigger. Revolving breech piece.

August 2, 1864, No. 43,709, **CHARLES ROBITAILLE** and **FLORIAN DAVIS**, Brooklyn, NY. Wedge lock open top, solid frame, single action, round butt revolver, front loaded with metallic cartridges having central teat at rear and split flared necks.

August 16, 1864, No. 1,737, **ETHAN ALLEN**, Worcester, MA. Reissue of No. 21,400 guard features.

August 16, 1864, No. 1,738, **ETHAN ALLEN**, Worcester, MA. Reissue of No. 21,400, anti-fouling features.

September 6, 1864, No. 44,126, **WILLIAM TILESTON**, Georgetown, DC. Set screw to be placed in lower part of frame and to engage recesses in cylinder to keep same from moving when not in use and to keep hammer between loaded chambers.

September 20, 1864, No. 44,303, **SIVE GUILBERT**, New York, NY. Sword handle and revolving firearm combined.

September 20, 1864, No. 44,363, **STEVEN W. WOOD**, Cornwall, NY. Barrel locks to frame above and below cylinder in front and swings on cylinder pin to permit loading and ejecting. Front loading. Tapered metallic cartridge cases, held in place by retaining plates at front or rear of cylinder (two systems). Rear of chamber closed save for small aperture for hammer nose.

October 24, 1864, No. 44,784, **ROBERT J. COLVIN**, Lancaster, PA. Gun and pistol bayonets combined.

October 25, 1864, No. 1,802, **ROLLIN WHITE**, Lowell, MA. (Reissue.)

November 8, 1864, No. 44,953, **FREEMAN W. HOOD**, Worcester, MA. (R.N. Eddy, attorney.) Fixed ejector as integral with cylinder pin. Cartridges seated in tubes which may be drawn forward against ejector to expell shell. Six shot, single action, rim fire, sheath trigger, resembling Allen and Wheelock stock.

November 22, 1864, No. 45,176, **HENRY REYNOLDS**, New Haven, CT. Pivoted extractor for single shells located on right side of frame behind cylinder at three o'clock in single action, solid top, solid frame revolver, for teat cartridges. (Pushes cartridges forward.)

November 29, 1864, No. 45,290, **ROLLIN WHITE**, Lowell, MA. Single action, solid frame, solid top, Center hammer, square butt, with trigger at front end of frame, and shouldered stock to permit better grip. Front loading teat cartridges.

December 20, 1864, No. 1,839, **A.P. SLOCUM**, Brooklyn, NY. Reissue of No. 38,204.

December 20, 1864, No. 45,532, **EBEN T. STARR**, New York, NY. Hammer lock incorporated in cocking spur to prevent premature discharge through slipping of hammer etc. (Adapted to Starr Army revolver.)

January 17, 1865, No. 45,912, **WILLIAM C. DODGE**, Washington, DC. Barrel rotates to left on bolt in front end of frame below a fixed cylinder. Top dovetail lock in front of hammer. Metallic cartridge, single action, rim fire.

January 24, 1865, No. 45,983, **WILLIAM C. DODGE**, Washington, DC. As above, but barrel rotates on a vertical axis at front end of frame below cylinder. Another like above, but with barrel rotating to right.

January 24, 1865, No. 46,023, **REUBEN H. PLASS**, New York, NY. Five shot, single action, rim fire revolver with trigger guard. Stock resembles loose ammo of Colts. Cylinder revolves on short as well as long axis of arm, rear end swinging off on a hinge to permit loading and extracting.

January 31, 1865, No. 46,131, **FREDERICK D. NEWBURY**, Albany & Catskill, NY. Single action, rim fire, solid top, solid frame, revolver for teat cartridges, front loading, rammer swings to side to seat cartridge in chamber at right side of frame. Rear ejector pushes shell out forward after firing.

February 7, 1865, No. 46,225, **WILLIAM H. ELLIOT**, Plattsburg, NY. Cylinder only shown. Discloses cylinder cap with cuts for hand and hammer nose as used in Remington conversions. This to remain fixed, and front part to be drawn forward for loading and extracting.

February 7, 1865, No. 46,243, **BENJAMIN W. JOSLYN**, Stonington, CT. (Henry Howson, attorney.) Single action, rim fire, revolver. Special type of cylinder lock involving vertical pin running through frame into indentations in rearward prolongation of cylinder and actuated by hammer nose. Barrel and frame held together by dovetail joints.

February 28, 1865, No. 46,562, **PATRICK HAUGHIAN**, New York, NY. Single action, rim fire, solid frame, solid top revolver with butt like Hammond deringer. Sheath trigger.

February 28, 1865, No. 46,612, **C.E. LUCINDER**. Typical open top Colt type single action revolver with cap magazine in butt. Chambers bored clear through cylinder, and breeches fitted with screw plugs to which nipples are attached.

April 4, 1865, No. 1,928, **ROLLIN WHITE**, Hartford, CT. Reissue of No. 19,961.

April 11, 1865, No. 47,252, **ALEXANDRE GUERRIERO**, Genoa, Italy. (A. Pollok, attorney.) Six shot, pin fire, double action revolver. Embodies a cylinder and breech cap, both sliding forward on cylinder pin, the cap actuated by the hand and imparting its rotation to the cylinder. Cap housed by a fixed breech cover. Sliding side ejector. Loaded by sliding barrel and cylinder forward.

April 18, 1865, No. 1,936, **WILLIAM H. ELLIOT**, Plattsburg, NY. Reissue of ring trigger pepperbox patents Nos. 32,378 and 33,382.

April 18, 1865, No. 1,937, **WILLIAM H. ELLIOT**, Plattsburg, NY. Reissue of ring trigger pepperbox patents Nos. 32,378 and 33,382.

April 18, 1865, No. 1,938, **WILLIAM H. ELLIOT**, Plattsburg, NY. Reissue of ring trigger pepperbox patents Nos. 32,378 and 33,382.

May 16, 1865, No. 47,707, **WILLIAM H. ELLIOT**, Plattsburg, NY. Improved cylinder pin for (Remington) revolvers.

May 16, 1865, No. 47,712, **GEORGE H. GARDNER**, New York, NY. Single action, solid frame, solid top, square butt revolver like Remington with two cylinders in line. Forward cylinder has five chambers with nipples, and one drilled all way through. Forward nipples fired by a rod working in top of frame when struck by the single hammer. When rear cylinder is to be used, the sixth chamber in forward cylinder is lined up, striking rod locked in forward position and hammer strikes nipples on rear cylinder.

May 16, 1865, No. 47,775, **JOHN H. VICKERS**, Worcester, MA. (Thomas H. Doge, attorney.) Six shot, single action, rim fire revolver, each chamber containing an inside sleeve or thimble into which the cartridge is loaded. An ejector rod fixed to a collar revolving around the barrel, lies parallel to barrel and thimbles may be drawn upon it to remove cartridges.

June 13, 1865, No. 48,201, **BENJAMIN F. PARKINSON**, Washington, PA. Revolving rifle. Large pulley inside butt, and small one in tang, over which a cord operates to feed cartridges into cylinder from a magazine in the stock.

June 20, 1865, No. 48,287, **BENJAMIN W. JOSLYN**, Stonington, CT. (Henry Howson, attorney.) Dovetail joints for attaching barrel to frame.

July 11, 1865, No. 48,775, **LOUIS C. RODIER**, Springfield, MA. Single action, rim fire, solid frame (type), solid top revolver with barrel and cylinder pivoted on a bolt in lower front end of frame in long axis of arm. Latch at top rear. Pushing on cylinder pin after gun is open ejects shells. (First use of hand ejector?)

October 3, 1865, No. 50,224, **SILAS CRISPIN**, New York, NY. Hinge below cylinder and midway of it. Latch vertically above. Front part of cylinder tips down with barrel, and cartridges are inserted therein, rear portions protruding and entering rear of cylinder when arm is closed. Hammer has long nose and strikes at junction of the two portions of cylinder.

October 3, 1865, No. 50,232, **WILLIAM H. ELLIOT**, Ilion, NY. Two vertically reciprocating firing pins with lock mechanism which causes each to serve one set of superposed barrels (four barrel pepperbox). Single hammer (concealed).

November 21, 1865, No. 51,092, **H. SMITH** and **D.B. WESSON**, Springfield, MA. Improved tip-up type.

November 21, 1865, No. 51,117, **WILLIAM MASON**, Ilion, NY. Cylinder rotates to right on horizontal axis at bottom of frame. Hand ejection. Remington type revolver.

November 28, 1865, No. 51,269, **JOSEPH RIDER**, Newark, OH. Attachments to adapt loose ammo arms for metallic cartridges or to be used in original manufacture of metallic arms.

December 12, 1865, No. 51,440, **WILLIAM H. ELLIOTT**, Ilion, NY. Many barreled firearm.

December 19, 1865, No. 51,628, **EBEN T. STARR**, Yonkers, NY. Starr, loose ammo revolver converted to take metallic cartridges.

December 26, 1865, No. 51,690, **GEORGE C. BUNSEN**, Belleville, NY. Revolving rifle, hammerless, with coiled mainspring and various gear trains in stock, operating revolving and firing mechanisms on release of trigger.

December 26, 1865, No. 51,762, **JAMES REID**, Catskill, NY. All metal knuckle duster, small caliber, pepperbox. Rim fire, single action, sheath trigger.

January 2, 1866, No. 51,836, **BENJAMIN W. JOSLYN**, Worcester, MA. Solid frame, open top, revolver with cylinder in two parts forming a chamber like that in Burnside rifle (powder case fits into rear portion, leaving bullet projecting. Front portion then slid back over bullets, barrel coming with it).

January 9, 1866, No. 51,985, **ELI WHITNEY**, New Haven, CT. (W.F. Lawrence, attorney.) A large caliber, rim fire, six shot revolver (looks like converted loose ammo) with breech cap into which the heads of shells are fitted. This is attached to a spindle concentric with another which runs through cylinder proper. By pulling back on cap (after removing it and cylinder from the arm by withdrawing cylinder pin), it travels far enough back to allow insertion or withdrawal of cartridges.

January 16, 1866, No. 52,105, **STEVEN W. WOOD**, Cornwall, NY. Single action, rim fire, solid top, solid frame, square butt, teat cartridge, revolver like tip up Smith and Wesson, with front loading cartridge tubes with split, flared necks, ejected by hook on side of frame at right below cylinder which passes through slots in side of cylinder to engage rear end of case.

January 23, 1886, No. 52,165, **HENRY HAMMOND**, Bridgeport, CT. Gun like Smith and Wesson tip-up, but latch at rear of cylinder below. Pulling up on a sliding collar which surrounds cylinder throws out the shells.

January 28, 1866, No. 52,248, **HENRY S. JOSSELYN**, Roxbury, MA. Single action, rim fire revolver with endless chain of steel links suspended for about 8 inches below frame in a loop and carrying twenty cartridges. Resembles bicycle chain. Each link acts as a chamber when passing through breech.

February 6, 1866, No. 52,504, **A. RAUB**.

February 13, 1866, No. 52,582, **BENJAMIN T. LOOMIS**, New York, NY. Seven shot, .22 (?) revolver with folding stock and trigger. Folds into very compact square shape. Double action, top hammer, circular main spring, and slotted cylinder in slot of which hammer ordinarily rests (safety feature).

March 27, 1866, No. 53,539, **WILLIAM R. MASON**, Ilion, NY. (Assd. to E. Remington & Sons.) Pivoted thumb piece at rear of cylinder withdraws shells singly at three o'clock (from Remington type single action, rim fire, solid frame, solid top, Army revolver.)

March 27, 1866, No. 53,548, **PROSPER POLAIN**, Brussels, Belgium. (A. Pollok, attorney.) A six shot, single action, rim fire revolver with trigger guard (Lefaucheux) having individual chamber tubes for the insertion of each cartridge. Extraction effected by drawing out the tubes.

March 27, 1866, No. 2,211, **J.W. COCHRAN** and **JOHN WEBSTER**, Enfield, NH. (Reissue.) Front loading cartridges of Burnside type, ramming and ejection carried out by linked rods and levers through rack and pinion.

April 10, 1866, No. 2,223, **JAMES WARNER**, Springfield, MA. (Reissue.) Shows receiver, solid top, solid frame, and barrel of revolving rifle.

April 10, 1866, No. 53,881, **SYLVESTER H. ROPER**, Amherst, MA. Revolving firearm.

April 17, 1866, No. 53,955, **GEORGE G. CROWELL**, Lime Rock, CT. Cylindrical (teat) shells are thrown out of chamber at three o'clock by forward motion of hammer actuating an ejecting mechanism on right side of frame. (Shape resembles Smith and Wesson tip-up.)

April 17, 1866, No. 54,065, **T.B. DOOLITTLE**, New Haven, CT. Frame and barrel dovetailed, seven shot, rim fire, single action revolver. Stock like tip-up Smith and Wesson. Barrel removes by twisting sideways and releasing dovetails. Cylinder cut away at rear so that bullets only enter it. A yoke actuated by the hammer fits over cartridge in axis of barrel at moment of discharge and thus completes the chamber. Ejection on cylinder pin fixed to bottom of barrel forging. (See patent July 29, 1862.)

May 1, 1866, No. 2,234, **HENRY REYNOLDS**, New Haven, CT. Reissue of No. 45,176.

May 8, 1866, No. 2,238, **S. CRISPIN**. Reissue of No. 50,224.

June 19, 1866, No. 55,743, **WILLIAM TIBBALS**, South Coventry, CT. Shows cylinder only and three forms of pin fire cartridges, one with long pin, one with short pin, and one with pin in center of base.

July 17, 1866, No. 56,466, **WILLIAM TIBBALS**, South Coventry, CT. Hinge and catch as in Smith and Wesson. For rimless rim fire cartridges like modern automatic ammunition. A ring block engages the slots in cartridge rims.

August 21, 1866, No. 57,448, **JOHN H. VICKERS**, Worcester, MA. Multi-barreled, rim fire, single action, pepperbox with collar surrounding barrels about midway and hinged at forward end of frame, below allowing them to tip down.

August 28, 1866, No. 57,622, **CHARLES A. CONVERSE** and **SAMUEL S. HOPKINS**, Norwich, CT. (Munn & Co., attorneys.) Pepperbox of Bacon type revolving within a collar which is held in place by a screw through bottom of frame. Single action, rim fire, sheath trigger. Side plate (left) and stock like tip-up Smith & Wesson.

September 11, 1866, No. 57,864, **ALBERT CHRIST**, California, OH. Hinge at front rear and catch at bottom, rear. Superposed barrels. Cylinder with two sets of chambers bored concentrically. Two hands (to insure revolution). One hammer.

September 11, 1866, No. 57,978, **F. SCHENCK**, San Antonio, TX. Hair trigger attachment for Colt loose ammo. revolver.

November 13, 1866, No. 59,629, **ALBERT L. MUNSON**, New Haven, CT. Hinge and catch as in Smith and Wesson. Cylinder has ratchets on both ends and is reversible. Insertion of cartridges ejects empty shells.

February 12, 1867, No. 2,480, **CHRISTIAN SHARPS**, Philadelphia (Harry Howson, witness.) Specification refers to Patent No. 22,753, January 25, 1859 (Reissue No. 1,199, June 18, 1861). Shows the Sharps 4 barrel pistol in various forms, with button barrel release, with a link mechanism to project barrels horizontally forward by throwing a guard lever forward and down, with the cartridges held in the fixed breech when barrels go forward, etc. Rotating hammer nose.

March 3, 1868, No. 75,016, **GEORGE HOLMAN**, Waterville, NY. (Munn & Co., attorneys.) A very long cylinder in a revolving rifle, alternate chambers being rifled and smooth, barrel smooth, and of larger diameter than cylinder to allow shot or bullet to spread.

April 2, 1867, No. 63,450, **REUBEN W. DREW**, Lowell, MA. Sliding side ejector and loading gate are attached to single action, rim fire, solid frame, solid top revolver of Smith and Wesson tip-up type.

June 4, 1867, No. 65,510 **ELISHA K. ROOT**, Hartford, CT. Cylinder pin is hinged below barrel and when released cylinder falls vertically and hangs like bunch of grapes. (No lower frame.) Cylinder locked in place by pin from rear.

June 11, 1867, No. 65,742, **EDWARD J. FROST**, New York, NY. A magazine (tubular) is fixed below barrel of a percussion type Colt, in place of bullet seater. It feeds flangeless cartridges into the chambers. Empty shells are driven out on right by attachment bolted to right side of hammer.

June 18, 1867, No. 2,650, **WILLIAM H. ELLIOT**, Plattsburg, NY. Reissue of Nos. 1936, 1937, and 1938.

July 9, 1867, No. 66,542, **ROLLIN WHITE**, Hartford, CT. Solid frame, open top, wedge lock, single action revolver with round butt. Takes front loaded, teat cartridges.

August 27, 1867, No. 68,292, **WILLIAM H. ELLIOT**, New York, NY. Single shot pistol with hammer acting as breech block as manufactured by Remington.

December 31, 1867, No. 72,844, **JOHN GORDON**, New London, CT. Spring power repeating firearm.

September 15, 1868, No. 82,258, **F. ALEXANDER THUER**, East Hartford, CT. Conversion apparatus for Colt revolvers.

September 22, 1868, No. 82,279, **CHARLES ETHAN BILLINGS**, Amherst, MA. Pistol and sword combined.

December 15, 1868, No. 84,976, **FRANKLIN WESSON**, Worcester, MA. Revolving firearm.

December 29, 1868, No. 85,350, **JOHN ADAMS**, Strand, Middlesex, England. Solid frame, closed top, revolver typically "Adams" with loading gate and sliding ejector on left side. Applicable according to the specification to conversion of muzzle loaders. For center fire cartridges.

May 4, 1869, No. 89,705, **LAFAYETTE Z. TERRILL**, Chicopee, MA. Repeating pistol with tubular magazine below barrel. Extractor hook on hammer. Square butt and sheath trigger. Five shot magazine.

July 20, 1869, No. 92,918, **FRANKLYN WESSON**, Worcester, MA. Pistol and dirk combined.

August 10, 1869, No. 93,572, **ROLLIN WHITE**, Lowell, MA. A spring actuated, axially located, ejector to eject all shells simultaneously from a tip-up type, single action, rim fire revolver.

August 10, 1869, No. 93,653, **ROLLIN WHITE**, Lowell, MA. Hand operated simultaneous ejector for tip-up type single action, rim fire revolver (lever under barrel).

August 24, 1869, No. 94,003, **CHARLES A. KING**, Springfield, MA. Single action, hinged frame, break open revolver with auto ejector operated by rack on cylinder pin and pinion ring around hinge. Square butt, Separate firing pin.

August 24, 1869, No. 94,035, **F. SCHENCK**, San Antonio, TX. Improved hair trigger for Colt loose ammo revolver.

August 24, 1869, No. 94,036, **F. SCHENCK**, San Antonio, TX. Hair trigger attachment to be fitted to trigger guard of single action revolvers of Colt type.

December 14, 1869, No. 97,780, **FRANCOIS ALEXANDRE LE MAT**, New Orleans, LA. A. Pollok, attorney.) Assigned to Charles Pietroni, London, England. Two barreled arm like the percussion Le Mat, but chambered for mettalic cartridges.

January 4, 1870, No. 98,529, **F. ALEXANDER THUER**, East Hartford, CT. Device to fit in frame of Colt and other revolvers to use rammer for seating bullets or caps on metallic shells.

February 1, 1870, No. 99,505, **ROLLIN WHITE**, Lowell, MA. Smith and Wesson type (hinge and catch same) with loading gate and spring-actuated ejector rod which is fixed to right side of frame, sliding through a guide.

February 8, 1870, No. 99,690, **JOHN M. MARLIN**, Hartford, CT. Extractor to remove shells singly through loading gate by hooking their flanges from the rear. Two forms. (For solid top, solid frame, single action, rim fire revolver.)

February 8, 1870, No. 99,693, **J.C. MILLER**, Revolving rifle with false butt plate separated from true by coil springs to absorb recoil. Side hammer.

February 15, 1870, No. 99,893, **B.B. HILL**, Cranston, RI. Cut shows cylinder only with ratchets at both ends and peculiar deep slots between chambers.

February 22, 1870, No. 100,227, **ROLLIN WHITE**, Lowell, MA. Superposed barrel revolver in two forms. A single shot barrel above revolver barrel in first case, fired by double nosed hammer. Upper barrel pivoted below to top of lower one, midway of it. Lower barrel hinged and latched like Smith and Wesson (can be released only when upper is swung sidewise). Second type has revolver barrel above, hinged like Smith and Wesson and single shot barrel below, hinged to tip down. Single shot barrel fired by lug on hammer striking cylinder pin.

May 10, 1870, No. 102,782, **CHARLES FELIX** and **JULES EDOUARD DE DARTEIN**, Strasbourg, France. (Munn & Co., attorneys.) Solid frame, open top, six shot revolver with cylinder revolved and hammer cocked by drawing back knurled knob in front of trigger, through cam action on curved lugs on surface of cylinder. Hammer has round knurled cocking piece. Front end of chamber rifled. Loading gate.

May 17, 1870, No. 103,013, **GEORGE W.H. CALVER**, Burlington, NJ. A hook ejector pivoted to hammer moves to rear as hammer goes forward, withdrawing the shell last fired (in single action, rim fire, solid frame revolver).

June 7, 1870, No. 104,087, **W.H. WERNER**. Cane gun.

June 21, 1870, No. 104,636, **WILLIAM J. PAGE**, Boston, MA. (E.P. Hale, attorney.) A hammerless six shot, ring trigger revolver. Solid frame with no projections. Loads by removing cylinder.

November 22, 1870, No. 109,417, **BENJAMIN W. JOSLYN**, Worcester, MA. Hinged extractor hook attached to hammer slips over flange of shell in axis of barrel and withdraws it when recoil shield is thrown aside (for Colt type open top, solid frame, single action, rim fire conversions).

November 22, 1870, No. 109, 513, **RANDALL D. HAY**. Safety device sliding vertically and interposing shield between hammer and nipple until trigger is pulled (for Remington Army type, loose ammo revolver.)

December 6, 1870, No. 109,914, **C.J. LINBERG** and **W.J. PHILLIPS**, St. Louis, MO. Open frame, open top revolver and revolving rifle with swing out cylinder and ratchets at front end of cylinder which is divided into two parts, front revolving and rear stationary. Through top of rear passes a firing pin which explodes cap on rear of front part when struck by hammer.

February 7, 1871, No. 111,534, **GILBERT H. HARRINGTON**, Worcester, MA. Sliding side ejector working through loading gate (for single action, rim fire, solid top, solid frame revolver.)

February 28, 1871, No. 112,127, **JOSHUA DAVIS**, Lime Stoneville, PA. Another revolving rifle with pulleys and cord in hollow stock feeding cartridges into cylinder.

March 7, 1871, No. 112,471, **EDWIN S. LEAYCROFT**, Brooklyn, NY. Teat cartridge revolver with front loading cylinder. Priming in tip of rounded base of cartridge.

March 7, 1871, No. 112,472, **EDWIN S. LEAYCROFT**, Brooklyn, NY. (Assd. to Colts) Hand operated single shell extractor (two forms) pivoted on right side of frame for ejecting front loading, metallic cartridges through loading gate.

March 28, 1871, No. 113,053, **SAMUEL S. HOPKINS**, Norwich, CT. Cylinder swings part way to right around horizontal axis in bottom of frame.

May 9, 1871, No. 114,653, **WILLIAM C. DODGE**, Washington, DC. Hinged and latched much like present Webley types. For center fire cartridges.

May 23, 1871, No. 115,258, **ELI WHITNEY**, New Haven, CT. (John E. Earle, attorney.) Type of cylinder pin lock as probably later used in the "Whitneyville" revolvers.

May 30, 1871, No. 115,483, **BENJAMIN W. JOSLYN**, New York, NY. Hinge at top rear. Latch at bottom rear. Joined ejector rod, housed in metal below barrel is manually operated by knob to extract shells singly. (Cylinder swings up with barrel.) Center fire type, single action, with separate firing pin.

June 13, 1871, No. 115,916, **FRANKLIN WESSON**, Worcester, MA. Widening out frame above and below cylinder so as to enclose more of cylinder surface.

June 20, 1871, No. 116,078, **JOHN L. MOSS** and **EDWARD W. JOHNSON**, Columbus, MS. Device for producing simultaneous extraction of all shells in Colt, open top, solid frame gun by an under lever action which causes barrel and cylinder to travel horizontally forward on cylinder pin, leaving shells sticking in fixed breech.

June 20, 1871, No. 116,225, **GEORGE W. SCHOFIELD**, U.S. Army. The Scholfield model, Smith and Wesson. Auto ejection by rack and pinion device.

June 27, 1871, No. 116,422, **SULLIVAN FOREHAND** and **HENRY C. WADSWORTH**, Worcester, MA.

July 4, 1871, No. 116,559, **F.G. COCHRAN**, St. Louis, Mo. (Munn & Co., attorneys.) Hand operated under lever produces simultaneous ejection in single action, rim fire, break open revolver shaped like Smith and Wesson tip -up. Forerunner of modern types.

July 4, 1871, No. 116,593, **FREEMAN W. HOOD**, Boston, MA. Cylinder swings to left. Shattuck style.

July 25, 1871, No. 4,483, **WILLIAM C. DODGE**, Washington, DC. (Reissue.) (Assd. to Horace Smith and Daniel B. Wesson, Springfield, MA.) Reissue of No. 45,912, Jan. 17, 1865. Barrel which is pivoted at front of and below cylinder, swings to left on an axis paralled to long axis of arm. Auto-extraction of shells. Crude drawings.

July 25, 1871, No. 117,461, **CHARLES B. RICHARDS**, Hartford, CT. (Assistant to Colt's Pat. Firearms Mfg. Co.) Mechanism for converting percussion Colts to center fire.

August 1, 1871, No. 4,496, **W.H. MORRIS** and **C.L. BROWN**, New York, NY. Reissue of No. 26,919.

August 15, 1871, No. 118,152, **JOSEPH RIDER**, Newark, OH. Five shot, tubular magazine pistol (magazine below barrel) with round butt and trigger guard. (Later manufactured by Remington with sheath trigger.)

September 5, 1871, No. 118,752, **CHRISTIAN SHARPS**, Philadelphia, PA. Barrel and cylinder swing to right on pivot in front end of frame (move in horizontal plane). Auto-ejection. Wedge latch at top rear. Dovetail latch in front of hammer.

September 19, 1871, No. 119,048, **CHARLES B. RICHARDS**, Hartford, CT. (Ass'd to Colt's.) The four shot Colt "Cloverleaf" revolver.

November 21, 1871, No. 121,199, **JACOB RUPERTUS**, Philadelphia, PA. (Howson & Son, attorneys.) Cylinder lock bites into cylinder at front end (at bottom of cylinder) being actuated by cam on hammer and spring in frame.

December 26, 1871, No. 122,182, **THOMAS LEE**, Westport, CT. Revolver for 41 rim fire cartridges having tubular magazine below barrel, which curves down below trigger guard and runs into butt, then upward and forward, the whole in shape of a question mark on its side, thus to feed cartridges one by one into a chamber on left side of cylinder.

April 9, 1872, No. 125,460, **ALBERT KARUTEZ**, Brooklyn, NY. Cane guns.

May 28, 1872, No. 127,377, **ANTHONY SMITH**, Hartford, CT.

June 25, 1872, No. 4,958, **ROLLIN WHITE**, Lowell, MA. (Reissue.) Hinge at top rear, latch at bottom front. Manually operated reciprocating cylinder pin ejects shells.

July 2, 1872, No. 128,644, **WILLIAM MASON**, Hartford, CT. (Ass'd. to Colts.) Refinements in sliding ejector mechanism as used in converting Colt open top, loose ammo arms.

July 16, 1872, No. 128,991, **DANIEL B. WESSON** and **CHARLES A. KING**, Springfield, MA. Refinements in auto ejector and cylinder catch for Smith and Wesson single action, break open, military type revolvers.

July 16, 1872, No. 129,334, **JOHN GORDON**, San Francisco, Calif. Solid frame, solid top, double action, revolving loose ammo rifle, Butt stock reciprocates on tang and by pressure on it gun is cocked and cylinder revolved.

October 1, 1872, No. 131,845, **EDMUND F. BURROWS**, Mystic River, CT. Safety hammer lock for single shot, tip down pistol of Stevens type.

October 22, 1872, No. 132,357, **JOSHUA [JARVIS] DAVIS**, Limestoneville, PA. (Chipmun, Hosmer & Co., attorneys.) Involves a moveable socket attached to barrel and fitting into recesses around the several chambers. Actuated by explosion gases, first to close and then to unlock the joint.

December 10, 1872, No. 133,732, **CHARLES S. WELLS**, New Haven, CT. Mechanism to be added to to Smith and Wesson tip-up revolver which, when struck by hammer, ejects the last fired cartridge shell through the loading groove on right. Also pictures a loading groove on left and a box magazine with a horizontal row of cartridges attached thereto, spring fed, and a hand actuated rammer by which cartridges may be pushed forward into a chamber one at a time.

December 31, 1872, No. 134,442, **OTTO SCHEELOCK**, Brooklyn, NY. Cylinder with triangular chambers whereby greatest number of chambers of a given size can be made in a cylinder of given diameter.

January 28, 1873, No. 135,377, **OTIS A. SMITH**, Middlefield, CT. (John E. Earle, attorney.) Single action, rim fire revolver with top cylinder stop resembling that on tip-up, Smith and Wesson side plate. Solid frame.

January 28, 1873, No. 135,378, **OTIS A. SMITH**, Middlefield, CT. (John E. Earle, attorney.) Hinge and latch as in Smith and Wesson tip-up, Pulling forward on cylinder pin tip releases cylinder as used later in "Smith Patent" revolvers.

February 25, 1873, No. 136,134, **WILLIAM CLEWS**, Ilion, NY. Single action, rim fire, solid frame, solid top, six shot, sheath trigger, round butt. Shell retractor hooks on head of shell in cylinder at 3 o'clock and withdraws it as hammer is raised to half cock.

February 25, 1873, No. 136,348, **DANIEL B. WESSON**, Springfield, MA. Frame below cylinder pivoted to that portion containing lock parts, so that rear portion may be revolved to right (latch at top rear). Barrel and cylinder also hinged at bottom front.

March 18, 1873, No. 136,871, **JOHN T. SMITH** and **JOSEPH J. SWEENEY** and **GEORGE M. SMITH**, Springfield, MA. Device to connect ejector at will with hammer (ejector oscillates in long axis of arm along right side of frame, withdrawing one shell at time through loading gate.)

March 18, 1873, No. 137,043, **DAVID WILLIAMSON**, Greenville, NJ. Break open revolver, auto ejector. Special lock construction and latch. Single action, rim fire, round butt.

April 15, 1873, No. 137,968, **OTIS A. SMITH**, Middlefield, CT. (John E. Earle, attorney.) Similar but new type cylinder stop.

April 22, 1873, No. 138,047, **GEORGE W. SCHOFIELD**, U.S. Army. New type extractor for Schofield model revolver. Device to keep gas out of lock mechanism.

April 22, 1873, No. 138,157, **GARDNER L. HOLT** and **JOSEPH C. MARSHALL**, Springfield, MA. (Ass'd. to Edward H. Buckland and Almanagor A. Buckland, Springfield MA.) Single shot, solid frame pistol with hammer as breech block, round butt, short barrel and trigger guard.

June 3, 1873, No. 139,461, **CHARLES FOEHL**, Philadelphia, PA. Single action, rim fire, tip-up revolver with hinge and latch like Smith & Wesson; round butt. Spring pawl on hammer acts on rear of cylinder shaft to revolve cylinder.

June 17, 1873, No. 140,028, **C.F. GALAND**, Paris, France. (John E. Earle, attorney.) A center fire, double action revolver with a (left) side plate hinged to frame forward of trigger. Trigger and main springs held by adjustable set screws (tension may be varied) within the stock.

July 1, 1873, No. 140,516, **JOHN M. MARLIN**, New Haven, CT. Single action, rim fire, tip-up revolver like Smith & Wesson. Round butt. Circular (left) side plate.

August 26, 1873, No. 142,175, **WILLIAM "H." PHILLIP** (Possibly "M.J.") Brooklyn, NY. Same principle as 47,712, but three cylinders in line instead of two. One chamber of each of rear cylinders contains a firing pin. Rotation by stud and grooves on cylinder faces.

September 2, 1873, No. 142,376, **B.K. DORWART**.

September 30, 1873, **ROLLIN WHITE**, Lowell, MA. New type of ejecting mechanism for break open, single action revolvers of Smith and Wesson type.

October 14, 1873, No. 143,566, **SULLIVAN FOREHAND** and **HENRY C. WADSWORTH**, Worcester, MA. Cylinder stop (on forward end of cylinder below) for a single action, rim fire, solid frame, solid type top revolver "Swamp Angel" type.

October 21, 1873, No. 143,855, **W. SIDNEY SMOOT**, Ilion, NY. (Ass'd to Remington. Dodge & Son, attorneys.) The regular Smoot model rimfire, single action revolver with round butt.

October 28, 1873, No. 5,631, **JACOB RUPERTUS**, Philadelphia, PA. (Howson & Son, attorneys.) Reissue of No. 43,606.

November 18, 1873, No. 144,814, **DAVID WILLIAMSON**, Greenville, NJ. Cylinder swings part way to right on pulling cylinder pin forward.

November 18, 1873, No. 144,815, **DAVID WILLIAMSON**, Greenville, NJ. Auto extractor made in three sections for two shells each, operating successively.

December 18, 1873, No. 5,691, **ROLLIN WHITE**, Lowell, MA. Reissue No. 143,394.

March 17, 1874, No. 148,742, **WILLIAM ORR**, New York, NY. Hinge and latch as in Smith and Wesson tip-up. Two cylinders. Hammer has a long nose, reaching to rear of front cylinder, and a short (adjustable) one serving rear cylinder. Disc between two cylinders serves as recoil shield for front cylinder and has passage in it for charges from rear.

March 31, 1874, No. 5,813, **ABRAM J. GIBSON**, Shrewsbury, MA. (reissue, ass'd to Daniel B. Wesson, Springfield, MA.) Hinge at top rear, latch at bottom front.

April 21, 1874, No. 150,120, **BENJAMIN H. WILLIAMS**, New York, NY. Cylinder travels forward on pin and leaves shells in fixed breech.

June 9, 1874, No. 151,882, **OWEN JONES**, Philadelphia, PA. (Howson & Sons, attorneys. John K. Rupertus witness.) Single action, rimfire, revolver with superposed barrels of different calibers revolving around a central axis, to be used with different cylinders, one kept in place and one housed in a recess in the stock under a pivoted cover.

September 15, 1874, No. 155,095, **WILLIAM MASON**, Hartford, CT. (Ass'd. to Colts. J.E. Earle, attorney.) The "New Model", pocket rim fire, single action Colt with loading gate.

December 15, 1874, No. 157,860, **DANIEL MOORE**, Brooklyn, NY. (J.P. Fitch, attorney. Ass'd. to Merwin, Hulbert & Co.) Six shot, single action, revolver with swing out mechanism like later Merwin & Hulbert guns.

January 19, 1875, No. 158,874, **DANIEL B. WESSON**, Springfield, MA. Improved auto ejector mechanism for single action or double action, break open revolvers.

January 19, 1875, No. 158,957, **WILLIAM MASON**, Hartford, CT. (Ass'd. to Colts. J.E. Earle, attorney.) Colt Frontier model, six shot, single action revolver.

February 23, 1875, No. 160,192, **FREEMAN W. HOOD**, Norwich, CT. (Ass'd. to Hood Firearms Co. N.U. Eddy, attorney.) Small single action rim fire revolver with sheath trigger and round butt. Solid frame. Embodies special cylinder pin catch. Ejection by cylinder pin. Side plate. Typical of its date.

March 9, 1875, No. 160,551, **DEXTER SMITH**, Springfield, MA. Auto extractor mechanism (cam action) for tip-up revolver of Smith and Wesson type.

March 23, 1875, No. 161,117, **DANA G. HADLEY**, Bethlehem, NH. Locking lever set below mainspring in Smith and Wesson tip-up type revolver.

March 30, 1875, No. 161,343, **JOHN H. LESTER**, Niantic, CT. (Ass'd. to self and E.F. Morgan, New London, CT.) Locking pin fits into recesses in base of hammer in Smith and Wesson type, tip-up revolver.

April 6, 1875, No. 161,615, **FREEMAN W. HOOD**, Norwich, CT. (Ass'd to Hood Fire Arms Co.) Cylinder stop operating in notches at rear end of cylinder in single action, rim fire, solid frame, solid top, revolver with round butt.

April 20, 1875, No. 162,162, **SULLIVAN FOREHAND** and **HENRY C. WADSWORTH**, Worcester, MA. Cylinder stop device on the military (Russian) model, single action, center fire, solid top, solid frame revolver with square butt.

April 20, 1875, No. 162,208, **ROLLIN WHITE**, Lowell, MA. Hinge and latch as in Smith and Wesson tip-up, Cartridge heads are held in recoil shield, and cylinder is moved foreward by hand operated lever after barrel is swung up, leaving shells in shield.

April 27, 1875, No. 162,475, **HENRY H. HOPKINS**, Norwich, CT. Cylinder only. Has safety notches for hammer to fall on to avoid burring chamber walls.

April 27, 1875, No. 162,582, **FREDERICK M. SHINN**, Leroy, KS. A tubular magazine lever action, revolving rifle. Magazine feeds into cylinder and cartridge is then pushed from cylinder into barrel by forward motion of breech block.

May 4, 1875, No. 162,863, **DEXTER SMITH, CHARLES C.** and **JOSEPH C. MARSHALL**, Springfield, MA. Breechloading firearm.

May 11, 1875, No. 163,032, **DEXTER SMITH**, Springfield, MA. Improvement on above.

May 11, 1875, No. 163,036, **DANIEL B. WESSON**, Springfield, MA. Extractor attached to recoil shield. Barrel and cylinder have two motions, forward to extract shells, and up (break-open) for loading.

June 29, 1875, No. 165,098, **CHARLES W. HOPKINS**, Norwich, CT. (Ass'd. to Merwin Hulbert & Co., New York.) Locking device set in front face of hammer and operating in conjunction with cylinder pin in single action, rim fire, solid frame, solid top revolver.

July 6, 1875, No. 165,369, **JACOB RUPERTUS**, Philadelphia, PA. (Howson & Son, attorney.) Typical little single action, rim fire revolver with round butt. Shows refinements in cylinder lock, (still situated at front end) and in cylinder pin lock.

July 27, 1875, No. 166,173, **ROLLIN WHITE**, Lowell, MA. Hinge and latch as in Smith and Wesson. Solid cylinder with ratchets on either end, reversible, also a divided cylinder, one half reversible. Lock improvements.

August 3, 1875, No. 6,569, **BENJAMIN H. WILLIAMS**, New York, NY. (reissue) No. 150,120.

October 11, 1875, No. 168,549, **FRANK H. ALLEN**, Norwich, CT. Hammer locked by pulling cylinder pin forward (in single action, rim fire, solid frame, solid top revolver.)

October 11, 1875, No. 168,562, **WILLIAM H. ELLIOT**, New York, NY. A short barrelled, four shot, single action, rim fire revolver, with round butt exactly like Remington deringer. Loading gate on left. Barrel apparently comes off for loading.

October 26, 1875, No. 6,715, **CHARLES W. HOPKINS**, Norwich, CT. (Ass'd. one half to Merwin Hulbert & Co., New York.) Reissue of No. 165,098.

November 9, 1875, No. 169,848, **JACOB RUPERTUS**, Philadelphia, PA. (Howson & Son, attorneys.) New type cylinder stop, fitting into notches in rear edge of cylinder at bottom and a split cylinder pin (to restrict expansion).

November 16, 1875, No. 170,038, **GEORGE E. WILLIAMS**, San Francisco, CA. Double action, solid frame, solid top revolver with new type lock.

December 14, 1875, No. 171,059, **DEXTER SMITH**, Springfield, MA. Hammer has prong to catch the extractor plate and hold it while cylinder is drawn forward (in tip-up revolver of Smith and Wesson type).

December 28, 1875, No. 171,506, **OTIS F. COLE**, Norwich, CT. Typical old single action, rim fire revolver with round butt and loading slot on right, fitted with a shield which is located at cylinder barrel joint and conducts gas from explosion into shell last fired, ejecting it.

January 18, 1876, No. 172,243, **EDWARD P. BOARDMAN** and **ANDREW J. PEAVEY**, Lawrence, MA. (C.A. Shaw, attorney.) Hammerless revolver with folding trigger operating a piston above barrel. Held in palm. (Sold as the "Little All Right.")

January 25, 1876, No. 172,716, **LUCIUS DIMOCK**, Leeds, Ma. Adjustable firing pin for a single shot pistol which may be set to strike either center or rim of a cartridge.

March 14, 1876, No. 174,731, **FREEMAN W. HOOD**, Norwich, CT. (Ass'd. to Hood Fire Arms Co.) Typical little .22 single action, rim fire, with solid frame, solid top, sheath trigger and round butt.

March 21, 1876, No. 175,180, **THOMAS J. SLOAN**, Tuckahoe, NY. Double action, hammerless, solid frame, solid top, center fire revolver with reciprocating trigger. Butt like Colt double action, army, six shot, sliding rod ejector and loading gate on right side of frame.

March 28, 1876, No. 175,518, **EBEN T. STARR**, New York, NY. Single action, rim fire, solid top, solid frame, sheath trigger, square butt revolver shaped like Smith & Wesson tip-up, hammer having two threaded recess at forward end, into either of which firing pin may be screwed (for center-fire or rim-fire cartridges). Sliding ejector folded against right frame which swings forward on a yoke to enter front of chamber when desired.

April 11, 1876, No. 175,951, **G.W. EDDY**.

April 18, 1876, No. 176,412, **DEXTER SMITH, CHARLES C. MARSHALL** and **JOSEPH C. MARSHALL**, Springfield, MA. Sliding extractor on right side of frame removes shells singly through loading gate (single action, solid frame, solid top revolver). Moving handle revolves cylinder and brings shells against extractor head.

April 25, 1876, No. 176,488, **DEXTER SMITH** and **JOSEPH C. MARSHALL**, Springfield, MA. In a six shot, single action revolver and extractor is pivoted to hammer in such a way that as hammer falls, its tip engages under rim of cartridge at 2 o'clock. As hammer continues to fall extractor moves to rear and ejects shell.

May 23, 1876, No. 177,887, **WILLIAM A. RICHARDSON**, Worcester, Ma. (One half ass'd. to G.H. Harrington.) Method of attaching spring to frame by developing an arc shaped recess in both. (Single action, rim fire, solid top, sheath trigger, round butt revolver.)

May 30, 1876, No. 178,133, **SULLIVAN FOREHAND**, Worcester, MA. (Ass'd. to self and Henry C. Wadsworth.) Sliding side ejector of Colt type on right side of frame in single action, solid frame, solid top, military type revolver.

June 13, 1876, No. 178,824, **STEPHEN W. WOOD**, Cornwall, NY. Ejector made part of cylinder pin mechanism of single action Army Colt, revolution of cylinder compresses ejector spring and tends to throw a shell out of loading gate with considerable force.

June 20, 1876, No. 179,026, **OWEN JONES**, Philadelphia, PA. Single action, break open revolver hinged like modern type. Cylinder remains in frame when barrel drops and is drawn forward by hand leaving shells in fixed breech

June 20, 1876, No. 179,084, **ROLLIN WHITE**, Lowell, MA. Sliding rod ejector, part below barrel and part to right of frame, so devised as to cause one-sixth turn of cylinder each time it is withdrawn from a chamber.

July 4, 1876, No. 179,633, **ROLLIN WHITE**, Lowell, MA. (J.W. Ozer & Co., attorneys.) Revolver somewhat like the Galand. Under lever when thrown forward actuates link which pulls barrel and cylinder forward leaving shells in fixed breech. Double action, center fire. Shells also withdrawn singly by a hook on hammer.

July 18, 1876, No. 180,037, **BENJAMIN F. JOSLYN**, Worcester, MA. (Ass'd. to R.P. Bruff, New York City, Howson & Son, attorneys.) Six shot revolver with barrel which revolves on cylinder pin and is drawn forward with cylinder for extraction. Ejection by lever at forward end of frame pushing cylinder pin backward and extracting all loads at once by flanges at its rear end.

September 26, 1876, No. 182,646, **JOSHUA DAVIS**, Limestoneville, PA. Cylinder only. Embodies gas check device, cylinder passing forward and covering barrel at time of discharge through use of bottle neck ammunition. Retracting spring returns it later.

October 17, 1876, No. 7,355, **GEORGE W. SCHOFIELD**, U.S. Army. (Reissue) No. 138,047.

October 17, 1876, No. 183,389, **FRANK W. FREUND**, Cheyenne, WY. The Frontier Colt so modified that it may be disassembled and assembled without using tools.

October 31, 1876, No. 183,944, **BENJAMIN F. JOSLYN**, Worcester, MA. (Howson & Son, attorneys.) A single action, center fire revolver with fixed barrel and cylinder and a divided breech portion, hinged to right of frame and swinging over to expose breech of cylinder for loading and extracting. Hammer, sear, main spring and sear spring, as well as hand and hand spring, are housed in the moving portion of breech.

October 31, 1876, No. 183,993, **THOMAS M. WALLIS**, Philadelphia, PA. Revolving firearm

November 7, 1876, No. 184,145, **JOHN W. COCHRAN**, New Hamburg, NY. Eight shot, rim fire, single action pepperbox with square butt. Barrels move forward to load and extract.

January 2, 1877, No. 185,881, **ANDREW W. WHITMORE**, Ilion, NY. Single action, rim fire, solid frame revolver with round butt and cylinder catch operating in notches at top of cylinder on rear; catch lifted by lever pivoted to hammer. Cylinder retaining arrangement resembles that in "Smith's Patent" revolvers.

January 23, 1877, No. 186,445, **STEPHEN W. WOOD**, Cornwall, NY. Auto ejector.

January 23, 1877, No. 186,470, **HENRY L. GARDNER**, Springfield, MA. Channels cut diagonally from chamber to chamber permit gas of explosion to pass from the one where discharge is taking place, into the next, in which empty shell lies, driving same out through loading gate. (Single action, rim fire, solid frame, solid top revolver.)

January 23, 1877, No. 186,509, **DANIEL B. WESSON**, Springfield, MA. Improved lifter mechanism for auto extractor in Smith and Wesson break open type revolver.

February 6, 1877, No. 187,135, **BENJAMIN HAVILAND** and **GEORGE P. GURN**, Ilion, NY. Solid frame, single shot pistol with trigger guard, square butt, and hammer breech block.

February 13, 1877, No. 187,244, **WILLIAM G. AYRES** and **GEORGE WHITTAKER**, Brooklyn, NY. Hinge at top rear. Latch at front, center. Barrel only lifts up. Projection on rear of barrel moves cylinder forward on cylinder pin as barrel is raised, leaving shells gripped by fixed extractor.

February 20, 1877, No. 187,689, **DANIEL B. WESSON** and **JAMES H. BULLARD**, Springfield, MA. Improved cylinder latch and ejecting mechanism in Smith and Wesson, break open, single action revolver. (Ass'd. to D.B. Wesson.)

February 26, 1878, No. 200,794, **OWEN JONES**, Philadelphia, PA. Ejection on same principle as in 179,026. (Break open, single action revolver.)

March 6, 1877, No. 187,975, **WILLIAM A. HUBERT**, Brooklyn, NY. Merwin Hulbert type of revolver open top. Heads of cartridges held in fixed extractor when barrel and cylinder are moved forward.

March 6, 1877, No. 187,980, **DANIEL MOORE**, Brooklyn, NY. (C.H. Watson & Co., attorneys.) Barrel and cylinder revolve to right and pull forward on cylinder pin, effecting extraction. Single action, rim fire revolver with open top frame.

April 10, 1877, No. 189,305, **MARCELIN DAIGLE**, Houma, LA. Cane gun.

April 10, 1877, No. 189,360, **OWEN JONES**, Philadelphia, PA. Improvement on No. 179,026, cylinder being drawn forward by rack and pinion device, leaving shells in fixed breech.

April 17, 1877, No. 7,610, **DANIEL MOORE**, Brooklyn, NY. (Ass'd. to Merwin-Hulbert & Co., New York City, C.H. Watson & Co., attorneys.) Reissue of 157,860. Typical Merwin-Hulbert gun.

May 1, 1877, No. 190,240, **DANIEL MOORE**, Brooklyn, NY. Two types, first one hinged at bottom center with latch at top rear. Pinion gears cut in top of hinge engage a rack set in lower frame, lug on which projects up into groove on cylinder carrying it forward (and leaving shells in fixed extractor) as barrel tips down. In second type hinge is at top center and catch at bottom front. Pinion gears in hinge engage rack cut all way around cylinder (four grooves encircle it) thus moving cylinder forward.

May 8, 1877, No. 190,543, **JOHN BROOKS** and **THOMAS W. BEARCOCK**, Pittston, PA. (The latter ass'd. to F. Balcom, Pittston, Pa. Brooks not assignor.) Typical single action, rim fire, round butt, solid frame revolver. Looks like "Defender."

May 15, 1877, No. 190,825, **R.W. CHURCHILL**

May 22, 1877, No. 191,178, **C.H. RICHARDSON**, Philadelphia, PA. (Howson & Son, attorneys.) Revolver with two barrels and two cylinders, concentric, two hammers and triggers. Cylinders revolve independently. Extraction not explained.

July 17, 1877, No. 193,269, **DANIEL MOORE**, Brooklyn, NY. (Ass'd. to Merwin, Hulbert & Co.) Barrel and cylinder revolve part way to right. Dovetail catch just ahead of trigger.

July 24, 1877, No. 193,367, **SULLIVAN FOREHAND** and **HENRY C. WADSWORTH**, Worcester, MA. Double action, solid frame, rim fire revolver with curved butt and (left) side plate, later put out in single action type as "Swamp Angel".

July 31, 1877, No. 193,620, **GEORGE W. SCHOFIELD**, U.S. Army. Improved auto ejector in Schofield model revolver.

August 7, 1877, No. 193,836, **DEXTER SMITH**, Springfield, MA. (ass'd. to D.B. Wesson.) Improved auto ejector for break open revolver.

August 21, 1877, No. 194,489, **THOMAS M. WALLIS**, Philadelphia, PA. Revolving firearm.

August 29, 1877, No. 194,653, **JAMES A. CROCKER**, Providence, RI. Double barrel rifle with four shot chamber, two hammers, two triggers and finger lever which cocks hammer and rotates cylinder one half revolution to bring two chambers opposite the two barrels (which are placed side by side).

September 25, 1877, No. 195,562, **THOMAS W. BEARCOCK**, Pittson, PA. Cylinder stop notches cut in an annular flange at rear end of cylinder; stop located in bottom of frame. (Single action, rim fire, solid frame, solid top arm.)

October 23, 1877, No. 196,491, **DEXTER SMITH**, Spingfield, MA. (Ass'd. to D.B. Wesson.) Improved auto ejector for break open revolver (single action).

November 6, 1877, No. 196,794, **OSCAR FRANKENAN**, Nurnberg, Germany. Combined pocket book and pin fire revolver in quadrangular form. Folding trigger outside. Rest of mechanism inside. Cocking trigger uncovers a trap for bullet to issue from. Gun consists of cylinder only, no barrel.

November 21, 1877, No. 197,708, **RICHARD D. WILLIAMS**, Baltimore, MD. Typical single action, rim fire, solid frame revolver. Cylinder catch eccentric and located between each chamber, not opposite chambers. Ratchet capable of rotation apart from cylinder.

December 18, 1877, No. 198,228, **DANIEL B. WESSON** and **JAMES H. BULLARD**, Springfield, MA. (Ass'd. to S. & W.) In a Smith & Wesson, single action, break open revolver the stirrup serves as a lever, causing hammer to rebound by force of mainspring.

January 1, 1878, No. 198,745, **OWEN JONES** and **FRANK W. MARSTON**, Philadelphia, PA. In a tip-up revolver of Smith and Wesson type, the nose of hammer burrs the edge of the (rim fire) shell in falling so that the burr will overlie the extractor head. In opening arm, only fired shells are removed; the unfired, having no burr, slip by the extractor head.

February 26, 1878, No. 200,794, **OWEN JONES**, Philadelphia, PA. Barrel tips down (breaks open) and at same time carries cylinder horizontally forward far enough to leave shells sticking in fixed breech.

March 26, 1878, No. 201,855, **ROLLIN WHITE**, Lowell, Ma.

April 16, 1878, No. 202,350, **BENJAMIN F. JOSLYN**, Worcester, MA. (A. Partridge, witness.) Barrel and cylinder move horizontally forward, leaving shells in a split breech ring, vertically divided, which is manually opened and closed to grasp or release shell rims.

April 16, 1878, No. 202,351, **BENJAMIN F. JOSLYN**, Worcester, MA. (Ass'd. to D.B. Wesson.) Extractor of usual type in connection with cylinder pin. Extractor operates by pulling forward on barrel and cylinder after bringing barrel to side. Hand, not spring actuated.

April 16, 1878, No. 202,388, **D.B. WESSON**, Springfield, MA. Means to cause hammer to rebound to half-cock notch. (For single action, break open Smith & Wesson.)

April 16, 1878, No. 202,418, **AMEDEE T. DEMOUNCIE**, London, England. (Ass'd. one half to Michael Kaufman.) Solid frame, double action revolver with side rod ejector like French and Belgium types. Left side plate is hinged at forward end of frame below cylinder and swings out and forward in a vertical plane, exposing lock parts. Lower leaf of mainspring extends forward and acts as trigger spring and pawl spring. Rebounding mechanism.

April 16, 1878, No. 202,613, **ROLLIN WHITE**, Lowell, MA.

April 23, 1878, No. 202,627, **WILLIAM H. BLISS**, Norwich, CT. Typical single action, rim fire, solid frame gun with circular (left) side plate like "Liberty."

April 23, 1878, No. 202,915, **LUIS YBARRA**, Madrid, Spain. Automatic gas operated ejector device on left side of frame whereby gas from explosion is utilized to eject shell last fired. (Amount of gas used may be regulated.)

May 28, 1878, No. 204,334, **BENJAMIN F. JOSLYN**, Worcester, MA. (Ass'd. to D.B. Wesson.) Sliding extractor bolt on right side of frame is cammed forward against spring pressure by the revolution of cylinder. When cam trips, spring pressure throws a cartridge out of loading gate.

May 28, 1878, No. 204,335, **BENJAMIN F. JOSLYN**, Worcester, MA. (Ass'd. to Daniel B. Wesson, Springfield, MA.) Hinge at bottom front. Latch at center, rear. Barrel revolves on axis in long line of arm, swinging to left.

May 28, 1878, No. 204,336, **BENJAMIN F. JOSLYN**, Worcester, MA. (Ass'd. to D.B. Wesson.) Pivoted piece on hammer strikes lever at rear of cylinder to eject preceding shell when hammer falls. May be disengaged when desired (in double action, solid frame revolver with solid top).

May 28, 1878, No. 204,337, **BENJAMIN F. JOSLYN**, Worcester, MA. (Ass's. to D.B. Wesson.) Extractor bar on right side of frame is cammed by rotation of cylinder to eject one shell at a time.

June 4, 1878, No. 204,438, **IVER JOHNSON** and **MARTIN BYE**, Worcester, MA. Double action, solid frame, solid top revolver with special lock mechanism.

June 11, 1878, No. 204,768, **JOHN D. SLATE**, Bernardston, MA. Single shot pistol with barrel sliding horizontally forward and taking a special (permanent) metallic cartridge shell and separate bullets.

July 30, 1878, No. 8,354, **GEORGE W. SCHOFIELD**, U.S. Army. (Reissue) Various devices to keep cylinder pin from fouling, to eject shells, lock cylinder, etc. Reissue of No. 193,620.

August 20, 1878, No. 207,168, **H.M. ERSKINE**, Ilion, NY. An extracting mechanism for single action, solid frame, solid top revolver actuated by cocking hammer. Pries out and ejects the empty shell at 2 or 3 o'clock. Right side plate.

December 3, 1878, No. 210,506, **WILLIAM C. DODGE, Washington,** DC. Special lock details in a double action, break open revolver of modern type.

December 10, 1878, No. 210,725, **ALONZO L. SWEET**, Norwich, CT. (Ass'd. one half to Bacon Arms Co.) Cylinder stop at bottom of cylinder near rear in cheap single action, rim fire, solid frame, solid top revolver with round butt.

January 28, 1879, No. 211,743, **BENJAMIN F. JOSLYN**, Worcester, MA. Break open type revolver in which revolution of cylinder is effected by breaking open (cylinder makes one-sixth turn). Ejector pushes shell at 6 o'clock out of gun. Top frame bends down over rear of cylinder and covers top half of rear face.

February 25, 1879, No. 212,606, **IVER JOHNSON** and **MARTIN BYE**, Worcester, MA. Double action, solid frame, solid top revolver with cylinder stop acting at bottom of cylinder near rear end.

March 11, 1879, No. 213,221, **PAUL MAUSER**, Oberndorf, Germany. Six shot, single action, center fire revolver, hinged frame just in front of hammer to tip up for extraction and loading. Coiled main spring in frame above trigger; cylinder revolved by cam action on helical slots.

May 20, 1879, No. 215,507, **ROSWELL F. COOK** and **JOSEPH RIDER**, Ilion, NY. (One half ass'd. to E. Remington & Sons, by each.) Solid frame, double action, solid top, center fire revolver in which one pull on trigger cocks hammer and a second pull fires it. Hammer may also be cocked by hand.

May 27, 1879, No. 215,721, **GEORGE W. CILLEY**, Norwich, CT. (Ass'd. one half to Hopkins & Allen.) Details of hammer and hand in a solid frame revolver. (Henry H. Hopkins and F.H. Allen, witnesses.)

July 8, 1879, No. 217,218, **JAMES C. HODGES** and **ALFRED A. HULL**, Morristown, TN. Revolving rifle with 20 shot cylinder. Lock plate when thrown down and forward throws barrel forward for inserting cylinder.

July 15, 1879, No. 217,562, **DANIEL B. WESSON**, Springfield, MA. A tubular magazine revolving rifle in which the hammer drives the shell a short way forward in barrel at time of firing, sealing the joint. Cartridges pictured are rimless.

October 28, 1879, No. 221,000, **DEXTER SMITH**, Springfield, MA. Smith and Wesson type, tip-up revolver. After barrel is swung up, cylinder is pulled forward by hand, leaving shells in fixed extractor.

November 4, 1879, No. 221,145, **A. EDWARD BARTHEL**, Detroit, MI. Hammerless, solid frame, solid top, revolving shotgun. Double action, "T" shaped hammer.

November 4, 1879, No. 221,171, **ANDREW HYDE**, Hartfield, MA. Shattuck type, hand ejection.

November 25, 1879, No. 222,065, **JOHN M. MARLIN**, New Haven, Ct. Smith and Wesson type tip-up. Extractor is part of cylinder, but stays in place until cylinder is moved forward to a certain point, where a catch is tripped and extractor released and allowed to retract into cylinder which may then be removed.

November 25, 1879, No. 222,066, **JOHN M. MARLIN**, New Haven, CT. Improvement on No. 222,065.

November 25, 1879, No. 222,098, **FERNANDO Y. SUNDERLAND**, Thorntown, IN. Tubular magazine revolving rifle, cartridges forced into barrel by breech block.

December 2, 1879, No. 222,167, **D.B.**

WESSON, Springfield, MA. Revolving rifle with oscillating breech block which extracts fired shells when hammer is cocked (by hand).

December 2, 1879, No. 222,168, **DANIEL B. WESSON**, Springfield, MA. Double action, Smith and Wesson, break open revolver with two series of stops arranged on parallel lines and actuated by a pivoted double stop.

December 30, 1879, No. 223,101, **WILLIAM H. BELL**, Baltimore, MD. Fall of hammer operates extractor to remove one shell through loading gate. May be disengaged. (Single action, solid frame, solid top revolver.)

January 20, 1880, No. 223,645, **PETER A. HOLTER**, Worcester, MA. Cylinder stop at bottom of cylinder near rear end (in double action, solid frame, solid top revolver like the Iver Johnson model 1900.)

April 27, 1880, No. 226,923, **EMILE NAGANT** and **LEON NAGANT**, Liege, Belgium. Solid top, solid frame, side ejector, double action, center fire revolver with trigger guard compressing mainspring, sear spring, and trigger spring.

May 11, 1880, No. 227,449, **GEORGE W. SCHOFIELD**, U.S. Army. Improved latch, etc., for Schofield model Smith and Wesson.

May 11, 1880, No. 227,481, **JAMES H. BULLARD**, Springfield, MA. (Ass'd. to D.B. Wesson.) Improved auto ejector for break open revolvers.

June 15, 1880, No. 228,859, **WILLIAM H. BELL**, Sr., Baltimore, MD. Side ejector on right side of a revolver like the single action Army, which is connected with cylinder pin so as to revolve cylinder one-sixth turn each time it is used to eject a shell.

June 15, 1880, No. 9,261, **BENJAMIN H. WILLIAMS**, New York, NY. Reissue of No. 150,120.

July 27, 1880, No. 230,582, **DEXTER SMITH**, Springfield, MA. Auto ejector which stays open until tripped by hand pressure on stud in hinge.

October 19, 1880, No. 233,363, **BENJAMIN MERRITT**, Newton, MA. (one half ass'd. to Melvin L. White, Boston.) Hammerless revolving shot gun for center fire cartridges.

October 19, 1880, No. 233,466, **JEAN MERINO BERGER**, St. Etienne, France. Ring trigger magazine pistol with tubular magazine below barrel holding three shots. Solid frame, one short barrel.

December 7, 1880, No. 235,240, **FREEMAN W. HOOD**, Norwich, CT. Superposed barrel, rim fire, single action revolver, closely resembling the "Duplex" (Osgood Gun Works). Axis of cylinder bored for a single large caliber cartridge, and for eight smaller caliber ones. Revolvable nose on hammer.

April 5, 1881, No. 239,634, **FRANK H. ALLEN**, Norwich, CT. (Ass'd. to Hopkins and Allen Mfg. Co.) Cylinder and pin removable to right, as in 1922 model. Baby Hammerless. Pressure on end of pin then ejects shells, also as in "Baby."

April 5, 1881, No. 239,821, **HENRY McGEE**, Norwich, CT. (Ass'd. to self and F.W. Hood.) Cylinder stop at bottom of cylinder near rear-end in single action, rim fire, solid top, solid frame revolver.

April 5, 1881, No. 239,914, **ROLAND L. BREWER**, Pittston, PA. (Ass'd. to J. Frank Lee, Wilkesbarre.) Cylinder pin lock for typical single action rim fire revolver.

May 10, 1881, No. 9,704, **ALONZO L. SWEET**, Norwich, CT. (Ass'd. to Bacon Arm Co.) Reissue of 210,725.

August 16, 1881, No. 245,792, **RANDOLPH P. CORY**, Union City, IN. (Munn & Co. attorneys.) Revolving rifle with cylinder housed in frame and rod hammer projecting to rear. Ring trigger draws cylinder back, revolves it and cocks hammer, then forces cylinder over end of barrel to form gas tight joint. This trigger is situated in front of firing trigger.

September 6, 1881, No. 246,817, **JULIUS N. PROESCHEL**, Milwaukee, WI. Folding (split) trigger and removable thumb piece (forms hammer) for solid top, solid frame, double action, center fire pocket revolver.

September 13, 1881, No. 246,940, **HOMER M. CALDWELL**, Springfield, MA. (Ass'd. one-half to Dexter Smith.) Double cylinder stop for double action, break open revolver evidently the Smith and Wesson frontier type.

September 20, 1881, No. 247,374, **WILLIAM MASON**, Hartford, CT. (Ass'd. to Colt's.) The Colt double action pocket and army model. No cylinder stop.

September 20, 1881, No. 247,217, **DEXTER SMITH**, Springfield, MA. Tip-up, single action revolver like Smith & Wesson. Cylinder moves forward on its pin, leaving shells caught on extractor located at head of cylinder pin. Pin may be removed and cartridges shaken out.

September 20, 1881, No. 247,218, **DEXTER SMITH**, Springfield, MA. Tip-up revolver with cylinder fixed to barrel lug, and auto extractor. Hinge at top rear and latch as in tip-up Smith and Wesson.

October 4, 1881, No. 247,938, **WILLIAM MASON**, Hartford, CT. (Ass'd. to Colts.) Details of lock for double action, solid frame revolver with cylinder turning anti-clock wise. Also a hammerless version.

October 11, 1881, No. 248,190, **WILLIAM MASON**, Hartford, CT. (Ass'd. to Colt's) Another double action, solid frame, type (revolution also anti-clockwise.)

October 11, 1881, No. 248,223, **DEXTER SMITH**, Springfield, MA. Latch at front, bottom, hinge at top, rear. (Smith and Wesson tip-up type.)

November 8, 1881, No. 249,240, **LYSANDER HILL** and **THERON S.E. DIXON**, Chicago, IL. Single action, tip-up like Smith & Wesson with thumb trigger on left side of frame.

November 15, 1881, No. 249,649, **WILLIAM MASON**, Hartford, CT. (Ass'd. to Colts.) Cylinder swings to right through horizontal axis in bottom of frame, then turns forward for ejection by a cam action on center pin.

December 6, 1881, **WILLIAM MASON**, (No. 250,375) Hartford, CT. (Ass'd. to Colts.) Like 249,624, but ejection occurs during last part of outward movement of cylinder and the secondary rotation is not necessary.

December 6, 1881, No. 250,591, **DEXTER SMITH**, Springfield, MA. Tip-up revolver like Smith and Wesson working on same principle as Marlin's No. 222,065. Extractor is held by hammer nose and released by half cocking latter.

December 13, 1881, No. 9,965, **DEXTER SMITH**, Springfield, MA. Reissue of No. 247,218. Single action, tip-up, auto ejecting revolver with hinge at top rear and latch at bottom front.

December 20, 1881, No. 251,306, **OTIS A. SMITH** and **JOHN T. SMITH**, Middlefield, CT. (Ass'd. to O.A. Smith.) Single action, rim fire revolver of break open type, with latch as now in common use. Non ejecting. Separate firing pin. (Latch is attached to frame, not to barrel.) Blow of hammer transmitted through vertical arm of latch to firing pin.

January 3, 1882, No. 251,750, **JOEPH H. WESSON**, Springfield, MA. Lock details of double action, Smith & Wesson break open type.

March 14, 1882, No. 254,798, **GEORGE W. GILLEY**, Norwich, CT. Mainspring made to act as trigger and pawl spring through intermediate lever pivoted to hammer.

March 21, 1882, No. 255,241, **EDOUARD BLED**, Paris, France, and **JEAN WARNANT**, Liege, Belgium. Details of lock for solid frame, double action, solid top, center fire revolver.

March 28, 1882, No. 255,573, **EDOUARD BLED**, Paris, France. Details of lock for solid frame, double action, solid top, center fire revolver.

July 25, 1882, No. 261,554, **MICHAEL KAUFMANN**, London, England. Two lugs on trigger act in a double set of notches in a double action, solid top, solid frame revolver of Belgian type.

July 25, 1882, No. 261,555, **MICHAEL KAUFMAN**, London, England. Details of lock for solid frame, double action, solid top, center fire revolver.

August 29, 1882, No. 263,551, **WILLIAM MASON**, Hartford, CT. (Ass'd. to Colt.) Lock details for solid frame, solid top, double action revolver. Also design for swing out cylinder release.

September 5, 1882, No. 263,684, **GEORGE W. CILLEY**, Norwich, CT. Lock details for double action revolver. Much like No. 254,798.

September 12, 1882, No. 264,325, **DANIEL MOORE**, Brooklyn, NY.

December 5, 1882, No. 10,257, **F.W. HOOD**, Boston, MA. (Reissue.) Reissue of No. 116,593.

December 5, 1882, No. 268,489, **FREEMAN W. HOOD**, Norwich, CT. Cylinder stop at bottom of cylinder near rear in single action, solid frame type revolver.

January 2, 1883, No. 269,890, **DONALD D. SMITH**, Springfield, MA. Single action, break open, revolver. Hammer has projection which hooks over rear of barrel extension when it falls, making an extra locking device.

January 9, 1883, No. 270,204, **GEORGE W. CILLEY**, Norwich, CT. Improved double action revolver lock.

March 6, 1883, No. 273,282, **ANDREW HYDE**, Hatfield, MA. The "Shattuck" revolver with cylinder and cylinder pin swinging out horizontally to right when released by catch where loading gate is usually located. Ejection on cylinder pin.

March 6, 1883, No. 273,644, **JACQUES E. TURBIAUX**, Paris, France. A circular palm revolver, fore runner of the "Protector," fired by pulling back on double hooks on either side of barrel (like pencil pistol); ten shots radially placed.

March 6, 1883, No. 72,335, **FRANK H. ALLEN**, Norwich, CT. Solid frame, solid top, double action revolver. Special lock devices.

March 13, 1883, No. 274,093, **HENRY GOODMAN**, St. Louis, MO. Cylinder swings to right parallel to barrel on axis attached to left side of barrel (in Colt double action type).

April 10, 1883, No. 275,377, **GEORGE W. HADLEY**, Chicopee Falls, MA. Hammer and firing pin quite like Maynard. Oscillating chamber piece for two cartridges, actuated by hammer to bring first one and then the other chamber in line with barrel.

June 12, 1883, No. 279,197, **DEXTER SMITH**, Springfield, MA. Barrel and cylinder swing to right around axis in bottom of frame. Wedge fastener in front of hammer. Auto ejection.

August 7, 1883, No. 282,552, **JOSEPH H. McLEAN**, St. Louis, MO. Six shot revolving rifle, double action, with extractor on left side of cylinder and loader on right, acting together through manual operation. A 68 shot steel box magazine attached to frame on right side in rear of cylinder.

August 14, 1883, No. 283,185, **JAMES T. ALDRICH**, Norwich, CT., (Ass'd. to self and Wm. H. Babcock.) The typical single action, rim fire, solid frame, solid top, "Liberty" revolver with notches at bottom rear of cylinder.

August 28, 1883, No. 283,854, **WILLIAM H. BLISS**, Norwich, CT. Hammer locking device sliding on top of frame. Solid frame, solid top, double action revolver with cylinder pin ejector. (Imitation break open.) Single spring serves hand and sear.

September 11, 1883, No. 284,786, **DANIEL B. WESSON**, Springfield, MA. Top hatch and cylinder retainer.

October 2, 1883, No. 285,862, **DANIEL B. WESSON**, Springfield, MA. Double action, hammerless revolver with separate firing pin and grip safety. Also same without safety. (Break open type.)

December 11, 1883, No. 289,956, **JAMES RANDALL**, Norwich, CT. Double action, solid frame, solid top revolver with cylinder pin ejection. Special lock mechanism.

December 11, 1883, No. 289,875, **DANIEL B. WESSON**, Springfield, MA. Grip safety for double action, break open hammer revolver.

January 1, 1884, No. 291,288, **ALEXANDER T. BROWN**, Syracuse, NY. Electric Gun Patents.

February 26, 1884, No. 294,188, **WILLIAM H. BLISS**, Norwich, CT. Double action, solid frame revolver resembling later types by Harrington & Richardson. Cylinder concaved at base so that cartridge rim bears only where it is to be struck by hammer.

March 11, 1884, No. 294,950, **FRANK H. ALLEN**, Norwich, CT. Cylinder stop at bottom near rear in double action, solid frame revolvers.

March 11, 1884, No. 295,064, **JOHN T. SMITH**, Rockfall, CT. Improved auto ejector and cylinder latch devices.

April 8, 1884, No. 296,325, **JAMES E. GAGE**, Concord, NH. Long barrelled, single shot, target pistol with breech block actuated by finger lever pivoted on left side of frame.

April 29, 1884, No. 297,801, **SAMUEL S. HOPKINS**, Norwich, CT. Sliding side ejector on right side of frame on a double action revolver. Ejection much like single action Army type.

July 1, 1884, No. 301,180, **WILLIAM TRABUE**, Louisville, KY. Cylinder swings to right parallel to barrel. Hand ejection.

July 1, 1884, No. 301,181, **WILLIAM TRABUE**, Louisville, KY. Improvement on No. 301,180.

July 1, 1884, No. 301,182, **WILLIAM TRABUE**, Louisville, KY. Double action revolver resembling later break open types, lower front portion of frame swinging laterally and carrying with it the cylinder to the right. Much like the present swing out cylinder types save that it throws to right instead of left. Ejection by pulling cylinder pin down.

July 8, 1884, No. 301,649, **JEAN WARNANT**, Liege, Belgium. Cylinder swings to right parallel to barrel. Hand ejection.

August 5, 1884, No. 303,135, **CARL T. EHBETS**, Hartford, CT. (Ass'd. to Colts.) Mechanism to release cylinder of single action Army Colt and allow it to swing to right on a pivot. Hand ejection.

August 5, 1884, No. 303,172, **HORACE LORD**, Hartford, CT. (Ass'd. to Colts.) Cylinder swings to left about an axis laid along top of frame at left side. Cylinder release at front end of frame.

August 19, 1884, No. 303,827, **CARL J. EHBETS**, Hartford, CT. (Ass'd. to Colts.) A swing out device (to right) almost identical with that of Lord (above). Hand ejection.

September 9, 1884, No. 304,731, **ARTHUR FRANCIS HOOD**, Hopkinson, MA. Double action, break open, center fire revolver with double indents between each chamber on cylinder surface and rear of cylinder milled circumferentially. This is without question the "American Arms Company" of Boston patent. Cylinder stop at bottom near rear.

September 30, 1884, No. 305,866, **DANIEL L. TOWER**, Brooklyn, NY. Double action, solid frame, solid top revolver with cylinder pin ejection. Special lock mechanism. Safety lock acting between trigger and hammer.

October 14, 1884, No. 306,596, **CARL J. EHBETS**, Hartford, CT. Lock parts for hammerless double action, solid top revolver like Colt double action Army.

October 21, 1884, No. 307,070, **SAMUEL RUSSELL**, Brooklyn, NY. Electric Gun Patents.

November 18, 1884, No. 308,183, **JOHN M. MARLIN**, New Haven, CT. Double action, break open, center fire revolver with cylinder stop at bottom near rear.

November 18, 1884, No. 308,231, **JAMES T. ALDRICH**, Norwich, CT. Double action, solid frame, solid top, rim fire revolver with imitation hinge. Cylinder stop at bottom near rear.

January 27, 1885, No. 311,323, **SAMUEL S. HOPKINS**, Norwich, CT. (Frank H. Allen, attorney.) Folding cocking spur for hammer on a double action revolver that looks like Merwin and Hulbert type.

January 27, 1885, No. 311,383, **JOHN T. SMITH**, Rockfall, CT. Lock parts for hammerless, double action, break open revolver of modern type.

March 3, 1885, No. 313,048, **WILLIAM H. BLISS**, Norwich, CT.

March 17, 1885, No. 314,067, **JOHN T. SMITH**, Rockfall, CT. Solid frame, (imitation break open) solid top, revolver. Double action. Special lock parts.

March 24, 1885, No. 314,494, **WILLIAM TRABUE**, Louisville, KY. Details of hand operated ejector for swing out cylinder revolver.

March 31, 1885, No. 314,754, **WILLIAM TRABUE**, Louisville, KY. Ejector plate with axis concentric with cylinder, capable of sufficient revolution to hook under flange of shells and eject same.

April 7, 1885, No. 315,352, **DEXTER SMITH**, Springfield, MA. Improved lifting device for auto ejector mechanism of double action, break open revolver.

May 12, 1885, No. 317,545, **FRANK E. KINSMAN**, Plainfield, NJ. Electric gun patents.

May 19, 1885, No. 317,965, **JAMES BOLAND**, Norwich, CT. Cylinder swings to right for ejection like Shattuck, then bends back parallel to frame for loading on a hinge set midway of cylinder pin. Hand ejection, Shattuck type.

May 19, 1885, No. 318,315, **DEXTER SMITH**, Springfield, MA. Improved lifting device for auto ejector mechanism of double action, break open revolver.

June 9, 1885, No. 319,898, **JOSEPH W. FROST**, Lawrence, MA. Electric gun patents.

June 30, 1885, No. 320,910, **W.C.A. BULLOCK**, Hinge and latch modern type. Barrel with cylinder ring about rear, forward end of cylinder being shouldered to revolve within ring.

August 4, 1885, No. 323,837, **DANIEL B. WESSON** and **JOSEPH H. WESSON**, Springfield, MA. Safety mechanism for Smith and Wesson break open hammerless revolver.

August 4, 1885, No. 323,838, **DANIEL B. WESSON**, Springfield, MA. New type grip safety for double action, break open hammerless revolver.

**August 4, 1885
to
January 11, 1887**

August 4, 1885, No. 323,839, **DANIEL B. WESSON**, Springfield, MA. Another new type grip safety for double action, break open, hammerless revolver.

August 4, 1885, No. 323,873, **JOHN S. LANDERS** and **D.B. WESSON**, Springfield, MA. (Ass'd. to Smith and Wesson.) Firing pin mechanism for Smith and Wesson hammerless, break open, double action revolver.

September 8, 1885, No. 325,878, **JOHN E. TYLER**, Roxobel, NC. (Munn & Co., attorney.) Cylinder only shown. Involves gas check for revolving arms.

November 3, 1885, No. 329,430, **EDWARD D. BEAN**, Boston, MA. Cane gun.

December 8, 1885, No. 332,071, **JAMES K.G. GAUITT**, Philadelphia, PA. Electric gun patents.

December 29, 1885, No. 333,416, **JOHN C. KELTON**, San Francisco, CA. Hammer safety operated by thumb latch on right side of frame (break open revolvers.)

January 5, 1886, No. 333,725, **JAMES BOLAND**, Norwich, CT. (Ass'd. to Hopkins and Allen.) (Frank H. Allen, attorney.) Hand device for solid top, double action, cylinder pin ejector, solid frame revolver.

February 9, 1886, No. 336,021, **JOHN T. SMITH**, Rockfall, CT. Lock details for double action revolver.

March 2, 1886, No. 336,894, **GEORGE W. CILLEY**, Norwich, CT. (Frank G. Parker, attorney.) Special lock parts and loading gate for double action, center fire revolver.

March 2, 1886, No. 337,201, **HENRY SCHLUND**, Aston Cross, Warwick Co., England. Hammerless, double action, center fire revolver. Modern type latch released by catch just above grip.

March 16, 1886, No. 337,872, **SAMUEL RUSSELL**, Brooklyn, NY. Electric gun patents.

March 16, 1886, No. 337,873, **SAMUEL RUSSELL**, Brooklyn, NY. Electric gun patents.

March 30, 1886, No. 338,760, **BENJAMIN MERRITT**, Newton, MA. (Crosby & Gregory, attorneys.) Hammerless revolving cylinder shotgun with gas check device and ring trigger. Loading gate at left.

April 6, 1886, No. 339,149, **GEORGE W. CILLEY**, Norwich, CT. Lock details for double action revolver. (Frame seems to be the Merwin and Hulbert type.)

April 6, 1886, No. 339,300, **IVER JOHNSON**, Worcester, MA. Lock details for double action, hammerless revolver of universal type.

April 6, 1886, 339,301, **IVER JOHNSON, R.T. TORKALSON** and **ANDREA FYRBERG**, Worcester, MA. Special lock mechanism, top latch, and ejecting device in double action, break open revolver.

April 6, 1886, No. 339,346, **REINHARD T. TORKALSON**, Worcester, MA. (Assd. to I. Johnson.) Lock details for double action, hammerless break open revolver.

April 6, 1886, No. 339,347, **REINHARD T. TORKALSON**, Worcester, MA. Improved details of auto ejector mechanism for double action, break open revolver. Special latch and ejecting devices, etc. A locking lever, curved like trigger, lies in trigger guard behind trigger. (Assd. to Iver Johnson.)

May 11, 1886, No. 341,696, **HUGH A. SILVER**, Cornhill, and **WALTER FLETCHER**, London, England. Opening loading gate automatically extracts the cartridge at two o'clock (in double action, solid frame, solid top revolver).

May 25, 1886, No. 342,507, **GEORGE H. FOX**, Boston, MA. Improved cam, etc., for auto ejector in double action, break open revolver.

July 20, 1886, No. 345,974, **IVER JOHNSON, R.T. TORKALSON** and **ANDREA FYRBERG**, Worcester, MA. (Latter two assd. to I. Johnson.) Complete double action, break open, hammerless center fire revolver. Numerous special points covered.

July 27, 1886, No. 346,327, **JOHN T. SMITH**, Rockfall, CT. Special type of cylinder catch and rebounding hammer in double action, break open, center fire revolver. (Separate firing pin.)

September 28, 1886, No. 349,864, **FRANK A. WARDWELL**, Methuen, MA. Cane gun.

October 5, 1886, No. 350,346, **GEORGE W. CILLEY**, Norwich, CT. (Assd. to Hopkins and Allen Mfg. Co.) Improved cam starter for auto ejector mechanism in double action, break open revolver.

October 5, 1886, No. 350,367, **CHARLES W. HOPKINS**, Norwich, CT. Improved cam starter for auto ejector mechanism in double action, break open revolver.

November 9, 1886, No. 352,185, **HENRY GOODMAN**, ST. Louis, MO. (Assd. to Daniel P. Kam.) Device applied to double-action Smith and Wesson break open revolver which eliminates pawl and trigger springs, acting through leverage of mainspring.

November 16, 1886, No. 352,863, **GUSTAF EUVALL**, Stockholm, Sweden. (Major, Swiss Army.) Double action revolver with cocking trigger in trigger guard and firing device on cocking spur of hammer. Promotes accuracy.

December 7, 1886, No. 353,914, **JAMES BOLAND**, Norwich, CT. Double action, center fire revolver with modern type latch. Hinge pin has eccentric lug in connection with auto ejecting mechanism. (Cut looks like Smith and Wesson in many ways.)

December 7, 1886, No. 353,948, **JOHN C. HOWE**, Worcester, MA. (Assd. to Sullivan Forehand.) Improved cam starter for auto ejector mechanism in double action, break open revolver.

December 28, 1886, No. 355,233, **DANIEL L. TOWER**, Brooklyn, NY. Improved cam-starter for auto-ejector mechanism in double action, break open revolver.

January 11, 1887, No. 355,761, **SULLIVAN FOREHAND**, Worcester, MA. Modern type, double action, break open revolver, but non-ejecting. Device for connecting cylinder with cylinder pin.

January 18, 1887, No. 356,387, **DANIEL B. WESSON**, Springfield, MA. New type of barrel latch.

February 1, 1887, No. 357,085, **IVER JOHNSON** and **ANDREW FYRBERG**, Worcester, MA. (Assd. to I.J.) Double action hammerless, break open, center fire revolver with separate hammer. Cylinder stop at bottom near rear and auto ejector. Details of lock and auto ejector mechanism for double action, hammerless, break open revolver with separate firing pin.

February 15, 1887, No. 357,710, **JOHN C. HOWE**, Worcester, MA. Modern type, double action, break open revolver. Cylinder released from pin by unscrewing.

February 15, 1887, No. 357,960, **WILLIAM L. HORNE**, Meriden, CT. Electric gun patents.

March 29, 1887, No. 360,263, **DANIEL B. WESSON**, Springfield, MA. New type of barrel latch.

April 5, 1887, No. 360,686, **GUILBERT H. HARRINGTON** and **WILLIAM A. RICHARDSON**, Worcester, MA. Safety hammer revolver as applied to "Young America" type of double action, solid frame, solid top, cylinder pin, ejector arm.

April 12, 1887, No. 361,100, **DANIEL B. WESSON**, Springfield, MA. Double action, break open revolver with double trigger guard (superposed) cocking trigger extending through to lower and firing trigger set in face of cocking trigger in upper.

April 12, 1887, No. 361,101, **DANIEL B. WESSON**, Springfield, MA. Wedge on top of breech portion of frame extends up through slot in rear end of barrel, extension to form barrel catch. Released by lateral pressure on spring latch.

April 26, 1887, No. 361,819, **FRANK A. FOSTER**, Norwich, CT. Improved cam starter for double action, break open, auto ejector revolver.

May 10, 1887, No. 362,631, **IVER JOHNSON** and **ANDREW FYRBERG**, Worcester, MA. (Latter ass'd. to former.) Double action, break open revolver showing special extractor mechanism, dust reduction devices, etc.

May 10, 1887, No. 362,632, **IVER JOHNSON** and **ANDREW FYRBERG**, Worcester, MA. (Fyrberg ass'd. to Johnson.) Double action, break open, hammerless revolver with various special details. Separate firing pin.

May 10, 1887, No. 362,637, **IVER JOHNSON** and **ANDREW FYBERG**, Worcester, MA.

July 5, 1887, No. 365,843, **EDGER A. MONFORT**, New York, NY. Electric gun patents.

July 12, 1887, No. 366,531, **DANIEL L. TOWER**, Brooklyn, NY. Folding finger rest to attach to lower rear left side of trigger guard of double action revolver to give support to middle finger when firing a pocket arm.

July 19, 1887, No. 366,794, **DAVID H. RICE**, Brookline, MA. Modern type, break open revolver. New method of securing cylinder to cylinder pin. Auto ejection.

August 2, 1887, No. 367,535, **JOHN M. MARLIN**, New Haven, CT. Cylinder stop at bottom near rear in double action, break open, center fire revolver.

August 9, 1887, No. 367,820, **JOHN M. MARLIN**, New Haven, CT. Modern type break open revolver. Cylinder retaining mechanism exactly like No. 366,794 at front end of cylinder, with additional special catch at rear end. Auto ejection.

August 9, 1887, No. 367,821, **JOHN M. MARLIN**, New Haven, CT. Improved cam starter for double action, break open, auto ejector revolver.

August 23, 1887, No. 368,599, **JOHN M. MARLIN**, New Haven, CT. New type barrel latch for double action, break open revolver.

October 4, 1887, No. 370,926, **HOMER M. CALDWELL**, Worcester, MA. (Ass'd. to Harrington & Richardson.) Modern type, break open revolver. New type cylinder release by pressing button on right or left side of top frame above middle of cylinder. Hand ejection by pushing cylinder pin to rear after gun is broken.

October 4, 1887, No. 370,964, **PAUL MAUSER**, Oberndorf, Germany. (Ass'd. to Waffenfabrik Mauser.) Repeating magazine (tubular, under barrel) pistol with ring trigger.

October 11, 1887, No. 371,532, **DANIEL B. WESSON**, Springfield, MA. Barrel latch.

October 18, 1887, No. 371,608, **JOHN M. MARLIN**, New Haven, CT. Barrel latch.

November 29, 1887, No. 373,893, **JOHN C. HOWE**, Worcester, MA. Double action, break open, hammerless revolver with top latch operated by catch at rear end of tang. Straight hanging stock (like auto) with round butt. Separate firing pin.

January 3, 1888, No. 375,799, **MARTIN BYE**, Worcester, MA. (Ass'd. to S. Forehand.) Double action hammerless, break open, revolver with whole top of rear frame lifting back to release barrel. Tang safety. Separate firing pin.

January 24, 1888, No. 376,922, **JOHN T. SMITH**, Rockfall, CT. Double action, hammerless, break open revolver with separate firing pin. Hammer may be set at full cock for single action work, or let down without snapping. Frame hinged to rear of tang and rear portion swings back and up to expose lock parts.

February 14, 1888, No. 377,877, **DANIEL B. WESSON**, Springfield, MA. New barrel latch.

February 14, 1888, No. 377,878, **DANIEL B. WESSON**, Springfield, MA. Another type of barrel latch.

February 21, 1888, No. 378,091, **AUGUST GRETH**, San Francisco, CA. Revolving rifle. Four shot.

March 13, 1888, No. 379,225, **IVER JOHNSON**, Worcester, MA. (Andrew Fyreberg ass'd. to Iver Johnson.) Different types of barrel latches.

March 13, 1888, No. 379,226, **IVER JOHNSON** and **ANDREW FYRBERG**, Worcester, MA. (Ass'd. to Iver Johnson.) Trigger locking device for double action, break open, center fire revolver.

March 13, 1888, No. 379,227, **IVER JOHNSON** and **ANDREW FYRBERG**, Worcester, MA. (Ass'd. to I.J.) A locking lever extending to rear of trigger guard must be pressed before trigger can be pulled back.

April 3, 1888, No. 380,361, **WILLIAM H. SPRAGUE**, Jamestown, NY. A hammerless, ring trigger revolver with curved barrel, curve beginning at chamber and ending midway of barrel, last one-half of which is straight. ("To make arm more compact, etc.").

April 10, 1888, No. 380,975, **EDWARD D. BEAN**, Boston, MA. Cane guns.

May 29, 1888, No. 383,701, **HOMER M. CALDWELL**, Worcester, MA. (Assd. to Harrington and Richardson.) Improved cam starter for double action, break open auto ejector revolver.

June 26, 1888, No. 385,009, **DAVID HALL KRICE**, Brookline, MA. (Ass'd. to Marline Firearms Company, New Haven, CT.) Improved cam starter for double action, break open, auto ejector revolver.

July 24, 1888, No. 386,535, **HENRY H. GRAHAM**, Chicago, IL. A freak rifle with drum magazine attached to lower surface of stock near toe, containing a belt of cartridges wire joined, which pass forward and up between two cylinders which mesh against the individual cartridges when they reach firing position. Water jacketed barrel.

October 16, 1888, No. 391,153, **IVER JOHNSON** and **ANDREW FYRBERG**, Worcester, MA. (Ass'd. to I.J.) Cylinder stop at bottom near rear in double action, break open revolver

October 16, 1888, No. 391,154, **IVER JOHNSON** and **ANDREW FYRBERG**, Worcester, MA. (Ass'd. to I.J.) Grip safety for double action, break open, hammerless revolver consisting of button at rear end of tang. Hammer safety. Barrel catch and cylinder lock.

October 16, 1888, No. 391,155, **IVER JOHNSON**, Worcester, MA. Breech loading firearm.

October 16, 1888, No. 391,156, **IVER JOHNSON** and **ANDREW FYRBERG**, Worcester, MA. (Ass'd to Iver Johnson.) Hammerless, auto ejecting revolver of modern type with wedge latch somewhat like Smith and Wesson hammerless. Cylinder released by pressing button on side of top frame. Separate firing pin. Special type of hammer rebounding device.

October 16, 1888, No. 391,213, **REINHARD T. TORKLESON**, Worcester, MA. (Ass'd. to Iver Johnson.) Wedge top latch with side spring release for a hammer revolver. Special cylinder stop device.

October 16, 1888, No. 391,214, **REINHARD T. TORKELSON**, Worcester, MA. (Ass'd. to Iver Johnson.) Cylinder stop and barrel latch.

October 23, 1888, No. 391,612, **ARTHUR F. HOOD**, Hopkinton, MA. (Ass'd. to American Arms Company, Boston, MA.) Cylinder stop at bottom near rear in double action, break open revolver. Stop pivoted below hammer.

October 23, 1888, No. 391,613, **ARTHUR F. HOOD**, Hopkinton,MA. (Ass'd. to John P. Lowell Arms Co.) Improved cam starter for double action, break open, auto ejector revolver.

November 6, 1888, No. 392,503, **CARL J. EHBETS**, Hartford, CT. (Ass'd. to Colts.) Swing out cylinder with rear thumb release of type later made by Colt. Hand ejection.

December 4, 1888, No. 394,027, **JOHN T. SMITH**, Rockfall, CT. New type of pawl mechanism. New cylinder holding device. New barrel latch and firing pin arrangement.

December 25, 1888, No. 395,119, **HOMER M. CALDWELL**, Worcester, MA. (Ass'd. to Harrington and Richardson.) Improved pawl for double action, break open, center fire revolver.

February 26, 1889, No. 398,595, **STEPHEN H. EMMENS**, London, England. Single shot pistol, smooth bored, with button trigger and special projectile.

April 9, 1889, No. 401,087, **DANIEL B. WESSON**, Springfield, MA. Reinforcing walls of cylinder notches with plates of hard steel to reduce wear.

April 30, 1889, No. 402,331, **MICHAEL KAUFMAN**, Greenwhich, England. Barrel latch.

April 30, 1889, No. 402,423, **CARL J. EHBETS**, Hartford, CT. (Ass'd. to Colts.) Lock mechanism for double action, center fire, swing out cylinder revolver, hammerless. Hammer has long hinged firing pin.

August 6, 1889, No. 408,457, **HOMER M. CALDWELL**, (Ass'd. to Harrington and Richardson.) Special pawl, hammer lifter, and trigger for narrow frames.

October 22, 1889, No. 413,197, **JOHN M. MARLIN**, New Haven, CT. Barrel latch.

October 29, 1889, No. 413,975, **JOHN T. SMITH**, Rock Fall, CT. Hammerless solid frame, double action, non-breaking revolver. Catch on tang at rear locks or unlocks hammer. Other special devices. Separate firing pin.

November 19, 1889, No. 415,444, **HOMER M. CALDWELL**, Worcester, MA. (Ass'd. to Harrington and Richardson.) Improved cam starter for double action, break open, auto ejector revolver.

December 17, 1889, No. 417,672, **CHARLES FOEHL**, Philadelphia, PA. Rebounding hammer with long pivoted nose, in a solid frame, solid top, double action revolver, hammerless, with cylinder pin ejector.

January 14, 1890, No. 419,412, **HOMER M. CALDWELL**, Worcester, MA. (Ass'd. to Harrington and Richardson.) Typical Harrington and Richardson with cylinder release by pressing button on left of top frame.

February 18, 1890, No. 421,798, **DANIEL B. WESSON**, Springfield, MA. Barrel latch.

March 11, 1890, No. 422,930, **GEORGE H. FOX** and **HENRY F. WHEELER**, Boston, MA. Double action, hammerless, break open revolver with hammer safety (on tang) made also in hammer type. Pulling on trigger may be used to cock hammer only, or to cock and trip it both. Separate firing pin.

April 22, 1890, No. 425,979, **HOMER M. CALDWELL**, Worcester, MA. (Ass'd. to Harrington and Richardson.) Lock and ejection improvements in double action, break open, hammer revolver.

April 22, 1890, No. 426,373, **JOHN FRICK**, Laramie, WY. Cane gun.

May 13, 1890, No. 427,833, **HOMER M. CALDWELL**, Worcester, MA. (Ass'd. to Harrington and Richardson.) Cylinder stop at bottom near rear, hammer lifter, hand and stop lever all pivoted on same pin (in double action, break open, center fire, revolver of Harrington and Richardson type.)

June 3, 1890, No. 429,397, **DANIEL B. WESSON**, Springfield, MA. Barrel latch and hammer safety combined.

June 17, 1890, No. 430,243, **HENRY F. WHEELER**, Boston, MA. Device to alter a double action hammerless revolver to single action for target practice; hammer safety lock, special barrel latch, etc.

July 15, 1890, No. 432,368, **JOHN CARTER** and **WILLIAM JOHN WHITNEY**, Birmingham, England. (Ass'd. to T.W. and Henry Webley.) Double action, hammerless, solid frame, solid top revolver. Special lock design.

July 15, 1890, No. 432,369, **JOHN CARTER** and **WILLIAM JOHN WHITNEY**, Birmingham, England. (Ass'd. as above.) (T.W. and Henry Webley.)

September 9, 1890, No. 436,062, **JOHN F. MANAHAN**, Lowell, MA. New type hammer locking device for Iver Johnson type break open, double action revolver.

January 20, 1891, No. 444,823, **CHARLES FOEHL** and **CHARLES A. WEEKS**, Philadelphia, PA. (Ellis Spear, attorney.) Special type cylinder pin release for solid frame revolver.

January 27, 1891, No. 445,511, **HENRY LOVE**, Creusot, France. Double action, solid top, solid frame revolver with shouldered cylinder and special lock type.

February 3, 1891, No. 445,880, **JULIAN WARNANT-CREON**, Hoignee-Cheratte, Belgium. Magazine pistol with tubular magazine below barrel. Hammerless. Frame resembles Martini. Trigger without guard. Magazine spring compressing knob on left side (for easy loading). loading).

February 24, 1891, No. 447,219, **CHARLES A. WEEKS** and **CHARLES FOEHL**, Philadelphia, PA. Solid top, solid frame, double action, hammerless revolver with separate firing pin.

June 2, 1891, No. 453,421, **HOMER M. CALDWELL**, Worcester, MA. (Chas. H. Burleigh, attorney.) Double action, solid frame revolver with cylinder pin catch of type later very common.

July 21, 1891, No. 456,179, **JOSEPH H. WESSON**, Springfield, MA. Combined barrel latch and hammer locking device.

August 4, 1891, No. 457,278, **ADAMANTIUS CORAY HOUSTON**, Pickaway, W. VA. A solid frame, solid top, hammerless revolver with revolving trigger having four arms, pulling on one of which revolves cylinder, then a bevel gear and causes firing pin to follow a reciprocating motion, striking primer and withdrawing.

September 1, 1891, No. 458,687, **HENRY F. WHEELER**, Boston, MA. Pivoted stop with two lugs works in two sets of notches near bottom rear of cylinder in double action, break open, center-fire, hammerless revolver with separate firing pin.

September 15, 1891, No. 11,191, **DANIEL B. WESSON**, Springfield, MA. Reissue of 323,838.

September 15, 1891, No. 11,192, **DANIEL B. WESSON**, Springfield, MA. Reissue of 323,839.

December 15, 1891, No. 465,179, **IVER JOHNSON** and **ANDREW FYRBERG**, Worcester, MA. Double action, break-open revolver (hammerless) with hammer safety on tang. Separate trigger.

January 12, 1892, No. 466,952, **GEORGE F. BROOKS**, Worcester, MA. (Ass'd. to Harrington and Richardson.) Lock details for double action, break open, center fire revolver.

January 12, 1892, No. 467,089, **HORACE B. FORBES**, Ogden, UT. Tubular magazine revolving rifle. Double action, loading mechanism actuated by rear-ward pull on trigger.

January 26, 1892, No. 467,558, **CARLO VON PECKER**, Vienna, Austria. Double action, center fire revolver, chamber bored with two concentric rings of six chambers each and superposed barrels. A revolving double firing pin permits firing one or two shots at a time. Solid frame with sliding ejector on right and loading gate.

February 2, 1892, No. 468,243, **CHARLES FOEHL** and **CHARLES A. WEEKS**, Philadelphia, PA. Double action, break open, hammerless revolver with rebounding mechanism. Separate firing pin.

February 23, 1892, No. 469,387, **ANDREW FYRBERG**, Worcester, MA. (not ass'd.) Combined sear and rebound catch for solid frame, solid top, double action, center fire revolver.

February 23, 1892, No. 469,465, **CARL J. EHBETS**, Hartford, CT. (Ass'd. to Colts.) Device to prevent accidental discharge in "New Army" type of Colt revolver.

March 22, 1892, No. 471,112, **CHARLES FOEHL** and **CHARLES WEEKS**, Philadelphia, PA. Barrel latch as later used in Foehl and Weeks revolvers.

April 19, 1892, No. 473,370, **THOMAS J. THORP**, Chicago, IL. Multishot tubular magazine gun with revolving cylinder and special cartridges with concave bases.

May 3, 1892, No. 473,903, **VICTOR BOVY**, New York, NY. Solid frame, solid top, double action, break open revolver, Six shots. Trigger not sheathed but without guard. Cylinder pin ejector. Means to effect gas-tight joint, etc.

June 7, 1892, No. 476,591, **ISRAEL N. ARCHBOLD**, Ridge Farm, IL. Revolving rifle using endless link chain, each link acting as a chamber.

July 18, 1893, No. 501,566, **HUBERT H. GRENFELL**, R.N., London, England and **JAMES G. ACCLES**, Birmingham, England. Solid frame, double action revolver with separate firing pin which moves in straight line.

August 29, 1893, No. 504,154, **PETER H. FINNEGAN**, Austin, IL. (Chaplin & Co., attorneys.) This is the Chicago Fire Arms Company "Protector." Seven shots.

September 26, 1893, No 505,569, **CHARLES M. HOPKINS** and **JAMES BOLAND**, Norwich, CT. Circular safety catch on tang of double action hammerless, break open revolver, which locks hammer when pushed forward and unlocks it by reverse movement.

October 3, 1893, No. 505,918, **IVER JOHNSON** and **ANDREW FYRBERG**, Worcester, MA. (ass'd. to Iver Johnson.) New type of top latch, and cylinder stop. (Wedge latch type like Smith and Wesson.)

November 21, 1893, No. 509,091, **JOHN L. McCULLOUGH**, Brooklyn, NY. Pump operated tubular magazine revolving rifle loaded with loose ammo and fired electrically. Hammerless.

November 28, 1893, No. 509,548, **JOHN L. McCULLOUGH**, Brooklyn, NY.

December 26, 1893, No. 511,406, **H.M. CALDWELL**, Worcester, MA. (not ass'd.) Improved lock for Harrington and Richardson type, break open revolver.

December 26, 1893, No. 511,620, **IVER JOHNSON**, Fitchburg, MA. (Oscar Mossberg ass'd. to Iver Johnson.) Barrel latch.

February 13, 1894, No. 514,490, **JOHN L. McCULLOUGH**, Brooklyn, NY. Hammerless, for metallic cartridges.

February 13, 1894, No. 544,696, **F. GEORGE P. BLOW**, Norfolk, VA. Hammerless revolver with cylinder swinging to right on a pivot at forward end of frame (moves in horizontal plane). Trigger and grip forward of cylinder so that gun fits hand like an automatic. Auto ejection.

March 13, 1894, No. 516,476, **ELBERT M. COUCH**, Northampton, MA. (Chalpin & Co., attorneys.) Improvement in actuating cylinder on No. 504,154. (No. 113)

March 27, 1894, No. 517,152, **DANIEL B. WESSON**, Springfield, MA. Cylinder swings to left on crane, being released by pulling forward on cylinder pin. Cylinder stop set into top of frame and lifted by falling and rising hammer.

April 3, 1894, No. 517,438, **ROSWELL F. COOK**, West Postdam, Ilion, NY. Cane gun.

April 17, 1894, No. 518,546, **O. JANKE**. Cane guns.

May 15, 1894, No. 519,875, **JOHN ROURKE**, Norwich, CT. Cylinder pin catch for single action revolver, old style, with round butt.

May 29, 1894, No. 520,468, **DANIEL B. WESSON**, Springfield, MA. Lock details in swing out cylinder Smith and Wesson. Each chamber surrounded at front end of cylinder by an annular projection, allowing combustion products to accumulate between.

July 3, 1894, No. 522,472, **JEAN LAMM**, Kesselstadt, (near Hanau) Germany. Tip-down (break open) revolver with cylinder held in place by two short pins at either end, a fixed rotating disc at breech bearing the ratchets and rotating cylinder. An extra supply of cylinders to be carried for continued fire. Cylinder is perfectly plain and can be loaded at either end; bears no cuts save the chambers.

July 10, 1894, No. 522,886, **E.E. DYBALL**. Cane guns.

August 2, 1894, No. 524,743, **CARL VON PECKER**, Vienna, Austria. double action, break open, superposed barrel revolver with large cylinder having two concentric sets of chambers of same caliber. One hammer and two firing pins.

October 30, 1894, No. 528,263, **OLIVER E. SMITH**, Chicopee Falls, MA. Improvements to "Protector" Palm revolver including loading indicator.

November 27, 1894, No. 529,877, **L.L.PICARD**. Revolver with knife attachment.

December 11, 1894, No. 530,730, **JOHN T. SMITH**, Rockfall, CT. (Earle & Seymour, attorneys.) Hammerless double action, solid frame revolver of cheap type with large left side plate (like later Colt and Smith & Wesson plates).

December 11, 1894, No. 530,759, **CHARLES FOEHL**, Philadelphia, PA. New type cylinder retaining mechanism in double action, break open revolver. (Also cylinder stop of new type).

March 5, 1895, No. 535,097, **FRANK B. FELTON**, Hartford, CT. (ass'd. to Colts) This is the Colt "New Army" revolver. Safety device incorporated.

April 2, 1895, No. 536,618, **ANDREW FYRBERG**, Worcester, MA. (Ass'd. to Harrington and Richardson.) Improved cam starter for double action, break open, auto ejector revolver.

May 21, 1895, No. 539,497, **DANIEL B. WESSON** and **JOSEPH H. WESSON**, Springfield, MA. Cylinder pin pushed forward by pressing button on left side of frame, thus releasing cylinder to rotate to left for ejection.

July 16, 1895, No. 542,744, **DANIEL B. WESSON**, Springfield, MA. Improved means of rotating cylinder and rebounding hammer in swing out cylinder Smith and Wesson with top cylinder catch.

July 16, 1895, No. 542,745, **DANIEL B. WESSON**, Springfield, MA. Two methods of keeping cylinder, when swinging out, from traveling to rear during ejection, one by a collar at base of cylinder rotating in a small slot in left side of frame, other by slot cut all around cylinder, concentric with its axis, and engaged by a stud on left side of frame.

October 1, 1895, No. 547,117, **NILS G. HANSON**. Cane guns.

October 8, 1895, No. 547,525, **FREDERICK SMITH**, Worcester, MA. (Ass'd. to Harrington and Richardson.) Improved cam starter for double action, break open, auto ejector revolver.

January 7, 1896, No. 552,699, **WILL-**

IAM A. **RICHARDSON**, Worcester, MA. Firing pin mechanism for double action, break open revolvers.

February 4, 1896, No. 554,058, **CHARLES FOEHL**, Philadelphia, PA. Stop lever pivoted at forward end of frame below cylinder. Catch lug works in notches at bottom near rear (in double action, break open, center fire revolver.)

March 17, 1896, No. 556,653, **JOHN T. SMITH**, Rockfall, CT. Cylinder stop lever pivoted to frame at top of cylinder in solid frame, solid top, hammerless (imitation break open) revolver with tang safety. Rear ward motion of hammer nose lifts catch.

April 7, 1896, No. 557,814, **GILBERT H. HARRINGTON** and **WILLIAM A. RICHARDSON**, Worcester, MA. (Ass'd. to H. & R.) Improved cam starter for double action, break open auto ejector revolver.

June 16, 1896, No. 561,963, **HOMER M. CALDWELL**, Worcester, MA. (Ass'd. to Mary E. Johnson, Fitchburg, MA.) The Iver Johnson hammerless revolver with removable tang cover. Separate firing pin.

August 4, 1896, No. 565,245, **DANIEL B. WESSON** and **JOSEPH H. WESSON**, Springfield, MA. Cylinder released by pulling forward on cylinder pin tip. Top cylinder catch.

August 4, 1896, No. 565,246, **JOSEPH H. WESSON**, Springfield, MA. Improved means of holding cylinder on cylinder pin.

August 11, 1896, No. 565,678, **CHARLES E. FOSTER**, Washington, DC. (Ass'd. to D.B. Wesson, Springfield, MA.) Hammer locking device for double action hammerless, break open, revolver operated by button on left side of frame.

August 11, 1896, No. 565,692, **WILLIAM A. RICHARDSON**, Worcester, MA. (Charles H. Burleigh, attorney.) Loading gate for double action solid frame revolver.

August 25, 1896, No. 566,393, **ANDREW FYRBERG**, Worcester, MA. (Ass'd. to Iver Johnson.) The safety cocking lever that made "hammer the hammer" famous. Applied to double action, solid frame, solid top revolver with separate firing pin.

August 25, 1896, No. 566,399, **IVER JOHNSON** and **ANDREW FYRBERG**, Worcester, MA. (Ass'd. to Iver Johnson.) New type cylinder catch.

December 22, 1896, No. 573,736, **DANIEL B. WESSON** and **JOSEPH H. WESSON**, Springfield, MA. Safety device for preventing discharge of swing out cylinder revolver until cylinder is locked in place.

June 8, 1897, No. 584,222, **JOHN H. HAMMER**, Allegheny, PA. Cane guns.

June 15, 1897, No. 584,631, **GEORGE V. FOSBERY**, London, England. Recoil operated revolver. Two forms. Both early types without helical channels on cylinder surface. One break open (like later model), and one solid frame.

June 15, 1897, No. 584,698, **JOHN L. MURPHY** and **DANIEL J. MANNING**, Springfield, MA. Cylinder locking device for swing out cylinder revolvers of Colt type.

December 7, 1897, No. 594,853, **BETHEL BURTON**, Brooklyn, NY. Revolving magazine bolt action, high power rifle. Clip loader, Seven shots.

March 8, 1898, No. 600,337, **WILLIAM A. RICHARDSON**, Worcester, MA. (Ass'd. to Harrington and Richardson.) (Mary A. Richardson, Executrix.) Safety lever somewhat like the above but separate from cocking lever.

April 26, 1898, No. 602,870, **GEORGE A. OWEN**, Springfield, MA. Auto ejecting mechanism for swing out cylinder revolvers of Smith & Wesson type.

June 28, 1898, No. 606,493, **WESLEY H. TRIPLETT**, Redlands, CA. Revolver with cylinder horizontally rotating, cartridges located radially. Twenty-four shots, double action hammerless. Round butt. Cylinder concaved slightly and running on ball bearing.

August 16, 1898, No. 609,233, **LEWIS T. HOUGHTON**, Worcester, MA. A cam shaped eccentric, pivoted transversely in front of hammer is partly rotated by cocking lever to permit hammer to fall full distance. Otherwise holds hammer away from firing pin.

September 13, 1898, No. 610,675, **ERIC ERICSON**, Florence, WI. Cane gun.

October 4, 1898, No. 611,826, **DANIEL B. WESSON** and **JOSEPH H. WESSON**, Springfield, MA. Cylinder stop at bottom near rear for Smith and Wesson swing out cylinder revolvers.

October 11, 1898, No. 612,071, **PERCY WAREHAM**, Brooklyn, MI. Improved device for securing auto ejection in break open revolvers.

November 29, 1898, No. 615,117, **DANIEL B. WESSON**, Springfield, MA. Two barrel latches, one at usual location on top rear of frame, and one at hinge making it impossible for an assailant to break open one's gun by grabbing top latch.

December 6, 1898, No. 615,467, **ARTHUR C. WRIGHT**, Worcester, MA. (ass'd. to Forehand Arms Co.) Cylinder stop a long hooked spring engaging at bottom of cylinder near rear (in double action, break open, center fire revolver.)

July 4, 1899, No. 627,966, **BURKHARD BEHR**, Bendlikon, Switzerland. Hammerless gun with two superposed barrels and a flat cylinder block with four chambers in a vertical line. After firing, cylinder is rotated on center pin, and the other two chambers come in line with barrels. Folding trigger. Large left side plate. Two firing pins fired alternately by same hammer.

July 4, 1899, No. 628,142, **ERIC ERICSON**, Florence, Wisconsin. Cane guns.

July 4, 1899, No. 628,356, **EDWARD S. RENWICK**, Millburn, NJ. Finger guard to run along right side of cylinder of revolvers.

August 8, 1899, No. 630,478, **BURKARD BELU**, Zurich, Switzerland. Four shot double barrel (superposed) revolver with chambers in vertical line, two being fired and other two then brought into allignment with barrels by revolving cylinder around its axis. Improved hammer mechanism over No. 627,966 q.v.

August 29, 1899, No. 631,951, **GEORGE W. CHANDLER**, Manhattan, KS. (Ass'd. one half to Wm. C. Snow, Boston, MA.) New lock device (involving deeper frame below cylinder) for swing out cylinder revolvers.

September 26, 1899, No. 633,734, **JACOB RUPERTUS**, Philadelphia, PA. Lock design for double action hammerless, break open revolver.

October 24, 1899, No. 635,705, **JOSEPH H. WESSON** and **JOHN L. HOBBS**, Springfield, MA. (Ass'd. to Smith and Wesson.) Mechanism to prevent cocking of swing out cylinder revolver when cylinder is swung out, or firing under same condition.

December 26, 1899, No. 640,070, **JACOB GOTTFRIED**, Aeschbacher, Rosario, Argentina. Subcaliber target barrel for break open revolvers (extra nose on hammer also.)

January 16, 1900, No. 641,620, **JOSEPH ASTER**, New York, NY. Foot operated revolver in shape of fish. Stepping on tail compresses trigger which is supported by two arms resting on floor, and discharges the piece. Releasing foot pressue permits trigger spring to act and arm to be recocked. To be used by military bands!

February 6, 1900, No. 642,688, **ANDREW FYRBERG**, Worcester, MA. Double action, hammerless, break open revolver with separate firing pin and special safety devices.

April 17, 1900, No. 647,861, **CHRISTOPHER D. MACDONALD** Vance, CO. Swing out cylinder, hand ejecting revolver, cylinder and barrel swinging to left in horizontal plane on pivot at forward end of frame.

May 15, 1900, No. 649,809, **HOMER M. CALDWELL**, Worcester, MA. (Ass'd. to Harrington and Richardson.) Hammer safety for double action, break open, center fire revolver whereby face of hammer presents a recess to rear end of firing pin until trigger is pulled, when a slide throws a solid piece opposite the firing pin and effects discharge.

June 5, 1900, No. 650,931, **CARL J. EHBETS** and **JAMES J. PEARD**, Hartford, CT. (Ass'd. to Colts.) Cylinder stop engages at bottom near rear and is actuated by one leaf of a two leaf main spring (in swing out cylinder revolver of Colt type.)

June 26, 1900, No. 625,625, **CHRISTOPHER D. McDONALD**, Vance, CO. Improved type of No. 647,861.

July 31, 1900, No. 654,850, **FREDERICK SHIELDS**, Erie, PA. Solid top, solid frame revolver with cylinder pin ejector. Rear ward pull on trigger cocks hammer and revolves cylinder. To trip hammer trigger must be released and allowed to move slightly forward.

August 14, 1900, No. 655,844, **JOSEPH H. WESSON**, Springfield, MA. Improved safety and rebounding features in Smith and Wesson swing out cylinder revolver of rim fire type.

September 18, 1900, No. 658,314, **MARTIN BYE**, Worcester, MA. (Ass'd. to Harrington and Richardson.) Device much like above and working on similar principles.

November 6, 1900, No. 661,450, **HENRY CLAY HALL**, Covington, KY. Toy revolver.

January 22, 1901, No. 666,476, **LEONARD HUNTRESS DYER**, Washington, DC. (Ass'd. to F.B. Pope, Augusta, Ga., and Jiles D. Ulims, Parkesville, S.D.) Hammerless revolver in which a spring actuated disc revolves when released by trigger and throws hammer against separate firing pin. Entire contents of cylinder discharged by single pull on trigger owing to power furnished by coil spring in disc.

January 22, 1901, No. 666,555, **FREDERICK B. POPE**, Augusta, GA. Revolving disc is wound up by act of "breaking" the arm. If trigger is held back all shots are fired; if released, but one. Grip safety on top of tang must operate with trigger to permit firing.

February 5, 1901, No. 667,248, **CHARLES P. O'NEILL**, Hartford, CT. Auto ejecting device added to Colt Army type swing out cylinder revolver.

March 12, 1901, No. 669,918, **FRANZ FORTELKA**, Vienna, Austria Hungary. (Ass'd. one half to W. Leitner.) Magazine pistol with box magazine in front of trigger. Horizontally oscillating round breech bolt is actuated by pull on trigger. Straight line hammer.

March 26, 1901, No. 670,862, **OSCAR F. MOSSBERG**, Fitchburg, MA. Cylinder stop engages at bottom near rear (is pivoted on top of trigger) in double action, break open, center fire revolver.

July 9, 1901, No. 678,274, **GEORGE A. OWEN**, Springfield, MA. Improved barrel catch and auto ejecting device for swing out cylinder revolvers. Cylinder swings out ninety degrees (to left).

July 29, 1901, No. 678,292, **GEORGE A. OWEN**, Springfield, MA. Revolving firearm.

August 13, 1901, No. 680,274, **JAMES J. PEARD**, Hartford, CT. (Ass'd. to Colts.) Means of locking hammer automatically in its rebounded position to prevent accidental discharge. (As adapted to Colt swing out cylinder type.)

September 10, 1901, No. 682,397, **DANIEL B. WESSON**, Springfield, MA. Safety bolt is interposed between hammer and frame when former rebounds, and remains in position until moved aside by action of hand in cocking cylinder.

October 8, 1901, No. 684,150, **DANIEL B. WESSON**, Springfield, MA. Rear cylinder pin lock for break open revolvers like those used in solid frame revolvers.

October 8, 1901, No. 684,331, **DANIEL B. WESSON**, Springfield, MA. Means to prevent friction between side of hammer and frame in Smith and Wesson swing out cylinder revolver.

October 29, 1901, No. 683,292, **CHARLES P. O'NEILL**, Hartford, CT. Improvement on No. 667,248.

November 5, 1901, No. 685,880, **AUDLEY H. STOW**, Thacker, W. VA. Type illustrated is like double action Colt but with break open mechanism. Barrel latch is on frame extension below cylinder at rear (like shot guns) with release button on lower left frame. Also a type with top and bottom barrel latches.

December 3, 1901, No. 688,141, **DANIEL B. WESSON**, Springfield, MA. Thumb cylinder pin release at left side of frame to rear of trigger. Forward end of cylinder pin housed under barrel.

December 3, 1901, No. 688,217, **WILLIAM J. WHITING**, Handsworth, England. Improvements on Webley-Fosbery auto revolver.

December 17, 1901, No. 689,260, **DANIEL B. WESSON**, Springfield, MA. Same cylinder pin release as above. Front cylinder pin catch as used in 1902 and 1905 Smith and Wesson models.

March 11, 1902, No. 694,969, **ARTHUR A. KEMP**, Loretta, OK. Finger guard for revolvers attached to loading gate and running forward parallel to cylinder wall opposite middle of right side of cylinder. Protects finger from gas.

April 1902, No. 696,539, **THEODORE J. BENNETT**, Orange, NJ. Ordinary .22 double action, solid frame revolver with a magazine attached above and to the left, feeding cartridges downward into a chamber on left side of cylinder.

April 15, 1902, No. 697,517, **OSCAR F. MOSSBERG**, Fitchburg, MA. (Ass'd. to Mary E. Johnson, Exec. of Iver Johnson.) Double action, hammerless revolver with cylinder swinging to right in horizontal plane, being pivoted at forward end of barrel near muzzle. Cylinder held in position by a cover which moves to right at right angle to axis of frame (not on arc of circle as does barrel.)

May 20, 1902, No. 700,592, **WILLIAM J. WHITING**, Handsworth, England. Cylinder stop for Webley-Fosbery auto revolver.

June 17, 1902, No. 702,607, **JOSEPH B. WESSON**, Springfield, MA. Improved means for locking cylinder in place in swing out cylinder revolvers.

June 17, 1902, No. 702,735, **HENRY M. KOLB** and **CHARLES FOEHL**, Philadelphia, PA. Double action, hand ejecting revolver, cylinder swinging to left and upward instead of down. Cylinder goes through two movements, first rotating outward on collar around barrel as an axis, then outward and up on the end of an arm pivoted to barrel as an axis.

July 15, 1902, No. 704,646, **FREDERICK H. JURY**, New York, NY. Cane gun.

September 2, 1902, No. 708,078, **JAMES D. ROBERTSON**, Norwich, CT. (Ass'd. to Thames Arms Co.) New type of cylinder catch (involving elasticity of a hollow split pin fixed in frame barrel and bearing an enlargement at its proximal end into which a corresponding enlargement on cylinder pin fits.) By pulling on cylinder pin it will slip out, owing to splitting of fixed hollow pin.

September 2, 1902, No. 708,437, **JOSEPH H. WESSON**, Springfield, MA. Safety catch at bottom of frame locks hammer in position away from face of cartridge. Catch is tripped by pulling back on trigger.

September 30, 1902, No. 710,008, **JAMES D. ROBERTSON**, Norwich, CT. (Ass'd. to Thames Arms Co.) Improved cam starter for double action, break open, auto ejector revolver.

October 7, 1902, No. 710,821, **GUSTAV TRESENREUTER**, Berlin, Germany. Revolver mechanism of small size suitable for enclosure in handle of umbrella or cane. Mechanism completely enclosed. Made in "T" shape with muzzle at one end of crossbar and rod hammer at other end. Folding trigger lets down below hammer.

December 30, 1902, No. 716,976, **ANDREW ALFSON**, Chicago, IL. Hammerless rifle with revolving cylinder, break open type, eight shot for bottle neck cartridges. Cylinder consists of two end discs with cartridge chambers hung between (What becomes of a chamber when cartridge is fired?) so as to allow free air circulation and rapid cooling. Barrel is surrounded by wood with air spaces all around, also to keep temperature down.

February 17, 1903, No. 720,775, **FRANK A. BECKWITH**, Evanston, WY. Safety device for single action Army Colt in connection with rear end of cylinder pin. Also spring and trigger improvements.

April 21, 1903, No. 726,109, **AUDLEY H. STOW**, Hunter, W. VA. Recoil operated, automatic revolver, breech block moving backward on explosion and cocking hammer, which revolves cylinder in usual way. Breech block operates like that of a lever operated Winchester. Same mechanism also adaptable to a tubular magazine pistol. Revolver is break open type.

May 26, 1903, No. 728,890, **ANDREW FYRBERG**, Worcester, MA. Top latch and cylinder catch.

June 30, 1903, No. 732,406, **MATT GOSS**, Duluth, MN. Revolving rifle with tubular magazine in stock, also cartridge reservoir feeding into magazine. Slotted cylinder surface.

July 7, 1903, No. 733,101, **DANIEL B. WESSON**, Springfield, MA. Improved combined trigger and hand spring for Smith and Wesson swing out cylinder revolver.

July 28, 1903, No. 734,524, **CARL J. EHBETS**, Hartford, CT. (Ass'd. to Colts; A.L. Ulrich, Witness.) Improved cylinder lock for swing out cylinder revolvers, etc.

August 4, 1903, No. 734,490, **ANDREW FYRBERG**, Worcester, MA. Top latch and cylinder catch.

October 20, 1903, No. 741,754, **MARTIN V.B. ALLEN**, New York, NY. (Ass'd. two-thirds to W.N. Thayer and Theodore Allen, New York City.) Lock for revolvers which can be operated only by one conversant with its design. (Old Smith and Wesson tip-up used as model.)

November 10, 1903, No. 734,784, **JOSEPH H. WESSON**, Springfield, MA. Mechanism for releasing cylinder to swing out by pressing forward end of cylinder pin from its catch. No rear release shown.

January 12, 1904, No. 749,212, **ALEXANDER L. MOLLETT**, Stonecoal, W. VA. Frame made in two part hinged at barrel hinge. Left part may be swung down and forward disc losing lock mechanism. Separate firing pin. Looks like Iver Johnson type.

January 26, 1904, No. 750,743, **WILLIAM J. WHITING**, Handsworth, England. Device to hold trigger stationary while hammer is being recocked by hand for single action target fire, and thus cause less disturbance of aim (in Webley break open type revolver.)

March 29, 1904, No. 756,182, **EUGENE NOVAK SZABADKA**, Austria-Hungary. Combined revolver and "knucks" of ingenious design, barrel situated between the two "knucks." Thumb operated trigger.

June 28, 1904, No. 763,581, **DANIEL J.B. WESSON**, Springfield, MA. Further improvements in combined trigger, and hand-spring for Smith and Wesson swing out cylinder revolver.

September 6, 1904, No. 769,277, **OTTO W. RINGQUIST**, Fitchburg, MA. (Ass'd. to M.E. Johnson.) Pivoted, rocking, cylinder stop, located at top of trigger, in double action, break open, center fire revolver.

November 8, 1904, No. 774,712, **OLE S. VOLD**, Dawson, MN. A combination hammer lock for double action, break open revolvers which cannot be released by persons not knowing the combination.

December 27, 1904, No. 778,500, **OSCAR F. MOSSBERG**, Chicopee Falls, MA. (Ass'd. to J. Stevens, A. & T. Co.) Hammerless revolver with grip safety and coil mainspring. Means of holding cylinder locked at all times save when actually being fired. Double action, break open, center fire, with separating firing pin.

December 27, 1904, No. 778,501, **OSCAR F. MOSSBERG**, Chicopee Falls, MA. Hammer type like above but without grip safety. Some cylinder lock devices.

April 11, 1905, No. 786,796, **LIONEL COBB**, Medford, MA. (Ass'd. to Mary E. Johnson, Fitchburg, MA.) Device for changing action of a hammerless break-open revolver from single action to double action by actuating a stop on left side of frame.

June 13, 1905, No. 792,077, **EDWARD S. RENWICK**, Millburn, NJ. Finger guard for revolvers. Situated parallel to cylinder at about 5 o'clock and concaved to fit finger. Extends beyond front end of cylinder.

June 13, 1905, No. 792,381, **MARTIN V.B. ALLEN**, New York, NY. Hammer lock which may be fitted to tang of any revolver by simply drilling two holes. Set and released by thumb. (Shown adapted to Harrington and Richardson break-open type.)

June 13, 1905, No. 792,382, **MARTIN V.B. ALLEN**, New York, NY. Another type of hammer lock, especially for double-action revolvers. Operator must understand mechanism to operate it.

July 4, 1905, No. 793,692, **GEORGE H. TANSLEY**, Hartford, CT. (Ass'd. to Colts.) Sliding lever blocks fall of hammer save when trigger is pulled.

August 1, 1905, No. 795,816, **MARTIN BYE**, Worcester, MA. (Ass'd. to Harrington and Richardson.) Hammer safety in tang operated by milled safety slide like shotgun safety.

February 6, 1906, No. 811,807, **JOSEPH H. WESSON**, Springfield, MA. Improved rebounding device for swing-out cylinder Smith and Wesson guns.

March 6, 1906, No. 814,017, **DUNCAN J. BUCHANAN**, Detroit, MI. Hammer cannot be cocked for single action work without first pulling slightly on trigger. (Foreign type of double action, solid top, solid frame, cylinder pin, ejector revolver shown.)

March 20, 1906, No. 815,490, **THOMAS E. MARTIN**, Batavia, NY. Electric gun patents.

March 27, 1906, No. 816,125, **JAMES D. ROBERTSON**, Norwich, CT. (Ass'd. to Thames Arms Co.) Starting cam improvement for ejector mechanism of double action, break open revolver.

April 17, 1906, No. 818,075, **MARTIN BYE**, Worcester, MA. (Ass'd. to Harrington and Richardson.) Pivoted stop catches at bottom of cylinder near rear.

April 17, 1906, No. 818,177, **HENRY M. KOLB** and **CHARLES FOEHL**, Philadelphia, PA. Cylinder swings out and up to right, rotating around two different axes in the process. Hand ejection.

April 24, 1906, No. 818,721, **JOSEPH H. WESSON**, Springfield, MA. New rebounding device embodied in a wedge block attached to hammer.

June 26, 1906, No. 824,197, **ELMER E. NEAL**, Springfield, MA. (Ass'd. to Smith and Wesson.) Top of hammer is hinged and lies against frame save when trigger is pulled.

July 24, 1906, No. 826,788, **HENRY M. KOLB** and **CHARLES FOEHL**, Philadelphia, PA. Means for locking cylinder pin in place when arm is cocked, etc.

August 21, 1906, No. 829,082, **JOHN J. MURPHY**, Norwich, CT. (Ass'd. to Hopkins and Allen.) Hammer is hung on an eccentric (?) and face strikes firing pin only when trigger is pulled.

November 6, 1906, No. 835,380, **JOSEPH H. WESSON** and **HARCOURT W. BULL**, Springfield, MA. (Not ass'd.) Improved rebounding mechanism for Smith and Wesson swing out cylinder revolver.

December 4, 1906, No. 837,867, **OSCAR F. MOSSBERG**, Chicopee Falls, MA. Four shot, squeeze type pistol with revolving hammer nose.

January 15, 1907, No. 841,240, **ANDREW FYRBERG**, Hopkinton, MA. Cylinder swings to right in horizontal plane. Auto ejector (Shattuck type.)

February 26, 1907, No. 844,671, **JOHN L. GARNER**, Washington, DC. Hammerless double action, break open revolver. Backward pull on trigger throws cylinder forward to make gas tight joint with barrel.

February 26, 1907, No. 845,274, **MAURICE STERN**, Philadelphia, PA. (Ass'd one-half to Louis Moskowitz.) Double action, swingout cylinder revolver with mainspring partly coiled and partly flat. All springs specially devised to permit maintenance of steady aim (in double action operation).

March 12, 1907, No. 847,011, **HENRY M. KOLB** and **CHARLES FOEHL**, Philadelphia, PA. Improvement on Nos. 818,177 and 826,788.

April 9, 1907, No. 849,825, **MARTIN VAN BUREN ALLEN**, New York, NY. Grip safety for swing out cylinder revolvers.

May 28, 1907, No. 855,227, **HARCOURT W. BULL**, Springfield, MA. New hammer retracting device for swing out cylinder type revolvers to block accidental discharge.

june 4, 1907, No. 855,439, **JOHN ADRIASON**, Chicago, IL. A Telescoping extension, perforated, to be attached to a revolver muzzle to reduce noise and furnish a handrest for long range firing.

June 4, 1907, No. 855,793, **JAMES C.E. LEACH**, Norwich, CT. Cut looks like Harrington and Richardson. New type of cam for lifting ejector rod in double-action, break open revolvers.

June 11, 1907, No. 856,286, **JOHN J. MURPHY**, Norwich, CT. (ass'd to Hopkins and Allen.) Hammer lock for hammerless, double action, break open revolvers, located on tang.

July 2, 1907, No. 859,032, **EARL T. ADAMS**, Portsmouth, OH. Cane gun.

July 16, 1907, No. 859,990. **MAURICE STERN**, Philadelphia, PA. Double action, break open type center fire revolver with coil springs.

August 20, 1907, No. 863,770, **WILLIAM J. WHITING**, Handsworth, England. Automatic pistol of conventional type.

December 31, 1907, No. 875,469, **JOSEPH TAMBOUR**, Nanterre, france. Hammer safety lock like the Iver Johnson, plus grip safety for double action, break open revolvers with separate firing pin.

January 7, 1908, No. 876,088, **E.O. PFEIL**. Electric gun patent.

February 11, 1908, No. 879,018, **WILLIAM STONE**, Lead, SD. Grip safety for Smith and Wesson hammerless revolver.

February 18, 1908, No. 879,295, **AUGUST MOELLER**, Chicago, IL. A twelve shot revolver with cylinder revolving in vertical axis of arm, provided with detachable bayonet and folding knife.

March 10, 1908, No. 881,374, **CHARLES C. CALL**, Springfield, MA. (Ass'd. to J.H. Wesson.) New trigger hammer mechanism is swing out cylinder revolver of Smith and Wesson type.

April 7, 1908, No. 884,119, **WALTER G. ZEISER**, Solomon, KA. Solid frame, double action revolver resembling Colt. Cylinder removable by pressing a spring release. Cartridge ejection as in 1924, model "Baby Hammerless." Ejector mechanism for swing out cylinder, hand ejector revolvers.

April 28, 1908, No. 886,211, **KEEMAZO HINO** and **TOMIJIRO KOMURO**, Tokyo, Japan. Automatic pistol with conventional lines. Reciprocating barrel, sheath trigger.

May 19, 1908, No. 887,784, **ANDREW FYRBERG**, Hopkinton, MA. Hammer safety. Separate firing pin.

May 26, 1908, No. 888,560, **JOSEPH C. WHITE**, Chelsea, MA. (Ass'd. to White-Merrill Company, Boston, MA.) Unusual automatic pistol of conventional appearance but capable of using barrels of different calibers and accommodating the suitable magazine in stock. May be used for single fire with automatic ejection. Side cocking feature for hammer, etc.

June 23, 1908, No. 891,510, **GEORGE H. TANSLEY**, Hartford, CT. (Ass'd. to Colts.) Safety device for automatic pistols.

June 23, 1908, No. 891,748, **JOSEPH TAMBOUR**, Nanterre, France. Safety device for Mauser military pistols and revolvers, operated by turning butt swivel through ninety degree arc.

July 14, 1908, No. 893,465, **ERNEST L. DURHAM**, Danville, VA. Device for compensating for wear in top latch.

September 1, 1908, No. 897,577, **GEORGE C. BOURNE**, Worcester, MA. Grip to be clamped at muzzle of a revolver to permit two handed shooting.

September 1, 1908, No. 897,806, **J.H. WESSON**, Springfield, MA. Firing pin held out of contact with primer by a light spring. No rebounding device needed as hammer lies in contact with firing pin but does not press it against primer save when it is purposely allowed to fall, firing pin working by inertia.

September 15, 1908, No. 898,717, **THOMAS F. BOWKER**, Fitchburg, MA. (Ass'd. to Mary E. Johnson.; Double action hammerless folding trigger revolver much like "Baby Hammerless."

November 17, 1908, No. 903,919, **R.T. TORKALSON**, Fitchburg, MA. (Ass'd. to Mary E. Johnson.) Double action, break open, center fire revolver. Special firing pin construction to resist battering action of hammer.

November 24, 1908, No. 905,020, **JOSEPH TAMBOUR**, Nanterre, France. Grip safety plus positive safety operated by sling swivel in butt. For double action, break open revolvers.

December 15, 1908, No. 907,069, **CHARLES H. MANN**, Delta, OH. Grip safety for hammer and hammerless revolvers released by pressing lever which projects through front of stock.

December 22, 1908, No. 907,670, **ARTHUR C. WRIGHT**, Norwich, CT. (Ass'd. to H. & A.) Stop pivoted above trigger, catches cylinder at bottom near rear.

January 5, 1909, No. 908,521, **ALGERNON P. TRUNDLE**, Perryville, AR. Automatic feed pistol with tubular magazine under barrel resembling magazine arms. Blow back type.

March 2, 1909, No. 913,756, **WILLIAM B. McCARTY**, Cale, IN. Revolver with cylinder rotating in vertical axis of arm, chambers drilled tangentially instead of radially and hammer and trigger mechanism being located in the hollow interior of the cylinder (ring) eighteen shots.

April 6, 1909, No. 917,045, **WILLIAM H. GATES**, Norwich, CT.

(Ass'd. to Hopkins and Allen.) Safety somewhat like the Iver Johnson but embodied in an arm separate from the cocking lever.

April 6, 1909, No. 917,723, **CARL J. EHBETS**, Hartford, CT. (Ass'd to Colts.) Safety located at right, rear, below and to right of hammer. There is both a firing pin lock and sear lock.

May 4, 1909, No. 920,303, **TERJE A. FIDJELAND**, Fostveldt, Norway. Automatic pistol (hammerless) of conventional type with cocking indicator.

May 4, 1908, No. 920,709, **E.M. ALDERMAN**, Sharon, PA. Electric gun patents.

August 3, 1909, No. 930,206, **PAUL MAUSER**, Oberndorf, Germany. Two types of automatic pistol, one pocket type and one larger, both conventional. Special safety features.

August 17, 1909, No. 931,146, **OTTO W. RINGQUIST**, Fitchburg, MA. (ass'd. to M.E. Johnson.) Cylinder catch located above trigger at bottom, near rear.

August 24, 1909, No. 932,183, **ANDREAS WILHELM SCHWARZLOSE**, Berlin, Germany. Automatic pistol in which barrel and slide both move forward against recoil spring at time of discharge.

September 7, 1909, No. 933,138, **EDWARD LEGGETT**, Brooklyn, NY. (Ass'd. to M.E. Johnson, Fitchburg, MA.) Improved type of Iver Johnson cocking lever safety.

September 14, 1909, No. 933,797, **CHARLES M. STONE**, Agawam, MA. (Ass'd. to Smith and Wesson.) Means of lessening tension required to cock hammer by retracting trigger in swing out cylinder revolver.

September 28, 1909, **ANDREW FYRBERG**, Hopkinton, MA New type of hammer safety for swing out cylinder revolvers. Separate firing pin.

October 5, 1909, No. 935,672, **THOMAS MARTIN**, St. Etienne, France. Automatic pistol of conventional design.

October 12, 1909, No. 936,369, **ELBERT H. SEARLE**, Philadelphia, PA. (Ass'd. three fourths to Wm. D. Condit, Des Moines, Iowa.) The Savage automatic pistol.

November 2, 1909, No. 939,111, **JOSEPH TAMBOUR**, Nanterre, France. Device for letting down hammer of automatic pistols without releasing safety.

November 9, 1909, No. 939,882, **WILLIAM J. WHITING**, Handsworth, England. Automatic pistol resembles Webley.

December 21, 1909, No. 943,819, **THOMAS M. DUBOSE**, Hattiesburg, MA. Revolver with superposed barrels of different calibers and one cylinder with two rings of chambers (five each) and a single firing pin actuated by one hammer, firing pin being movable vertically in a slot so as to function with either upper or lower barrel.

December 28, 1909, No. 944,448, **CHARLES F. LEFEVER**, Toledo, OH. Automatic recoil operated revolver with hammer. Barrel and cylinder recoil and cylinder turns by means of stud and slots like Webley-Fosbery. Break open type. Pictured in hammerless model also.

December 28, 1909, No. 944,606, **RUDOLF FROMMER**, Budapest, Austria-Hungary. Safety device for the Frommer automatic pistol.

Juanuary 4, 1910, No. 945,320, **ANDREW FYRBERG**, Hopkinton, MA. Safety device somewhat like the Iver Johnson (cocking lever, safety type) embodied in a solid frame revolver with cylinder which swings horizontally to left. Hand ejection.

January 11, 1910, No. 946,214, **WILLIAM H. GATES**, Norwich, CT. (Ass'd. to Hopkins and Allen.) Hammer rises out of firing position and its face engages frame as soon as trigger pressure is released.

January 11, 1910, No. 946,351, **GEORGES VANDER - HAGEN,** Liege, Belgium. Solid frame, recoil operated auto revolver, hammerless cylinder being revolved either by ratchet and pawl at front end, or by stud and slots as in Webley-Fosbery. Cylinder and breech bolt are the only recoiling parts. Barrel remains stationary.

January 25, 1910, No. 947,478, **JOHN M. BROWNING**, Ogden, UT. Thumb safety for automatic pistols of Browning type (to lock action either in open or closed position).

January 25, 1910, No. 947,481, **THOMAS CONSENTINO**, Houston, TX. (Ass'd. one half to C.J. Suhm and one-fourth to John P. De-Phillippe, Houston, Texas.) Automatic pistol dismountable by hand. Stock (containing magazine) at right angles to barrel.

February 15, 1910, No. 949,467, **RUDOLF FROMMER**, Budapest, Austria, Hungary. Locking device for automatic pistols.

April 5, 1910, No. 954,190, **HENRY M. KOLB**, Philadelphia, PA. The original "Baby Hammerless" revolver.

April 5, 1910, No. 954,191, **HENRY M. KOLB**, Philadelphia, PA. "Baby" hammerless. Laterally removable cylinder and special cylinder pin lock.

April 12, 1910, No. 954,657, **ABNER H. STOCKWELL**, Worcester, MA. (Ass'd. to Harrington and Richardson.) Sliding hammer nose is actuated by pulling trigger to line up with primer.

April 19, 1910, No. 955,436, **GEORGE B. REID**, Aberdeen, Washington. Double action, hammerless, break open revolver with separate firing pin. Discharge produced by releasing on trigger after hammer is cocked.

April 26, 1910, No. 956,430, **LOUIS SCHMEISSER**, Suhl, Germany. Automatic pistol with magazine in front of trigger and hinged barrel which tips down.

April 26, 1910, No. 956,431, **LOUIS SCHMEISSER**, Suhl, Germany. Automatic pistol of conventional type with magazine in stock behind trigger. Rod cocking indicator. Hammerless.

May 24, 1910, No. 959,229, **HENRY M. KOLB**, Philadelphia, PA. Double action, break open revolver with two pawls, one giving cylinder 1/10 rotation and next giving it another 1/10. When hammer rebounds after firing, No. 2 pawl rotates cylinder to a point where hammer rests one-half way between chambers.

June 14, 1910, No. 961,188, **JOSEPH H. WESSON**, Springfield, MA. Hammer safety actuated by the trigger through the hand.

June 14, 1910, No. 961,189, **JOSEPH H. WESSON**, Springfield, MA. Same as No. 961,188 with slight changes (stop not hand actuated.)

June 21, 1910, No. 962,137, **WILLIAM H. GATES**, New London, CT. Conventional type auto pistol with fixed barrel and reciprocating breech block.

July 12, 1910, No. 964,419, **NICOLO GIUSTO**, Italy, and **CARLO DE FELICE**, New York, NY. Revolving rifle with cartridge reservoir in butt and track conveyor which delivers them to cylinder.

August 30, 1910, No. 968,691, **OTTO W. RINGQUIST**, Fitchburg, MA. (Ass'd. to M.E. Johnson.) Cylinder catch located at bottom, near rear, in double action, break open hammerless revolver.

August 30, 1910, No. 969,011, **EDWARD B. VEDDER**, Ft. Stevens, OR. Subcaliber barrel for revolvers. Used with chamber adapters which permit employing gun without any alternation.

September 13, 1910, No. 970,248, **MYRON C. LISLE**, Toronto, Ont., Canada. Double action, hammerless revolver. Rearward motion of trigger throws cylinder forward to make gas-tight joint. Cylinder swings laterally to left in "Shattuck" fashion.

November 15, 1910, No. 975,685, **ANDREW FYRBERG**, Hopkinton, MA. Cylinder swings out and up to left, ending in about same position as Foehl and Weeks cylinder, but rotates about a single axis only (lying along left side of top of frame). Hand ejection.

December 6, 1910, No. 978,092, **JOSEPH H. WESSON**, Springfield, MA. Thumb actuated safety pivoted above trigger at its forward end and operating a pinion through rack at rear end, situated on left side of conventional automatic pistol. Must be squeezed at time of firing. Cannot be set beforehand.

December 13, 1910, No. 978,415, **JOSEPH H. WESSON**, Springfield, MA. Automatic pistol with hinge at top rear. Trigger guard hinged at top front (swings down and forward). Squeeze safety at front side of grip.

January 17, 1911, No. 982,150, **WALTER L. MARBLE**, Gladstone, MI. "Game Getter" with firing pins successively tripped by single trigger.

January 17, 1911, No. 982,152, **WALTER L. MARBLE**, Gladstone, MI. "Game Getter" with two nosed hammer.

February 14, 1911, No. 983,519, **JOHN M. BROWNING**, Ogden, UT. (C.J. Ehbets, attorney.) The Colt .45 automatic pistol.

February 28, 1911, No. 985,156, **RUDOLF FROMMER**, Budapest, Austria-Hungary. Locking device permitting semi-automatic or automatic fire according as trigger is held or released.

April 11, 1911, No. 989,242, **RUDOLF FROMMER**, Budapest, Austria Hungary. Improvement on No. 985,156.

April 18, 1911, No. 989,817, **HERMAN F. SIEGEL**, Brooklyn, NY. Safety lock to be attached to rear side of revolver trigger guard and block motion of trigger.

April 25, 1911, No. 990,669, **ISAAC H. RODEHAVER**, Auburn, CA. Superposed barrel, revolving rifle with one cylinder pierced with two sets of chambers of different calibers at different distances from axis, one set centering with each barrel. Hammer has two firing faces and is actuated by two triggers.

May 23, 1911, No. 993,129, **FRANK UNGER** and **WILLIAM CORRIGAN**, Cleveland, OH. Device to be worn about neck which will discharge two shots straight ahead if hands are raised as when surprised by robber.

June 13, 1911, No. 995,156, **HENRY M. KOLB**, Philadelphia, PA. Top latch for hammerless revolvers.

June 27, 1911, No. 996,252, **FREDERICK I. JOHNSON**, Fitchburg, MA. Improved type of Iver Johnson safety device.

August 1, 1911, No. 999,209, **LYMAN H. COBB**, Fitchburg, MA. (Ass'd. to Mary E. Johnson.) Cam improvement for ejector mechanism of double action, break open revolver.

September 26, 1911, No. 1,004,172, **FREDERICK I. JOHNSON**, Fitchburg, MA. Coil mainspring for Iver Johnson type break open revolver. Tension adjustable.

November 7, 1911, No. 1,007,709, **EDWARD B. LEGGETT** and **THOMAS W. APPLEGARD**, Brooklyn, NY. Hammerless double action, folding trigger revolver with new top latch and cylinder catch. Coiled mainspring.

March 5, 1912, No. 1,019,446, **CHARLES FOEHL**, Philadelphia, PA. Device actuated by hammer for automatically ejecting the last fired shell by hooking it out through loading gate, each time trigger is pulled (on Baby Hammerless type revolver.)

April 30, 1912, No. 1,024,932, **GEORGES VANDER - HAGEN,** Liege, Belgium. Means of setting safety lock in automatic pistols when magazine is withdrawn.

April 30, 1912, No. 1,024,933, **GEORGES VANDER - HAGEN,** Liege, Belgium. Another means of setting safety lock in automatic pistols when magazine is withdrawn.

September 3, 1912, No. 1,307,544, **FREDERICK T. RUSSELL**, Philadelphia, PA. Grip safties for Savage auto on front and back in handle. (Front blocks trigger; rear sear.)

October 15, 1912, No. 1,041,648, **PAUL MAUSER**, Oberndorf, Germany. Automatic pistol like later pocket models. Safety blocks sear lever until bolt is closed.

October 22, 1912, No. 1,042,007, **WILLIAM W. KEY**, Fayetteville, AR. Hammerless double action, break-open revolver with straight line action. Grip like automatic and below cylinder, not behind it. Space for extra cartridges in hinged hollow stock.

October 22, 1912, No. 1,042,145, **ISAAC H. RODEHAVER**, East Auburn, CA. Superposed barrel revolving rifle with two firing pins set in breech to be struck by firing faces of hammer. Single trigger. Eight shots large caliber and one shoots small. Firing pin to be struck is brought into operation through a manual adjustment.

December 3, 1912, No. 1,046,268, **LYMAN H. COBB**, Fitchburg, MA. (Ass'd. to Mary E. Johnson.) Improved coiled mainspring of adjustable tension for Iver Johnson break open revolvers.

December 10, 1912, No. 1,047,054, **BRADFORD B. HOLMES**, So. Bellingham, WA. A palm type revolver with hinged cover giving a flush exterior. Fired by palm pressure, barrel projecting between fingers. Looks somewhat like a cigarette case with barrel projecting.

December 17, 1912, No. 1,047,672, **PAUL MAUSER**, Oberndorf, Germany. Firing pin safety for automatic pistols.

December 31, 1912, No. 1,049,105, **WILLIAM W. KEY**, Fayetteville, KA. Same as No. 1,049,106 and 1,049,107.

December 31, 1912, No. 1,049,106, **WILLIAM W. KEY**, Fayetteville, KS. Same as 1,049,105 and 1,049,107.

December 31, 1912, No. 1,049,107, **WILLIAM W. KEY**, Fayetteville, KS. A revolver with covered hammer mechanism (hinged cover) and stock placed below cylinder to give good balance. Break open type. Trigger mechanism much like that of automatics. Side of grip (which is hollow) is hinged to permit the carrying of extra cartridges therein.

April 22, 1913, No. 1,059,581, **CARL WALTHER**, Zella St. Bl. Thuringia, Germany. Recoil operated auto pistol with stationary barrel.

May 20, 1913, No. 1,062,121, **HANS SAUER**, Suhl, Germany. (Ass'd. to T.P. Sauer & Son.) Trigger locking device for automatic pistols. Operates when magazine is removed and stays locked until released.

July 8, 1913, No. 1,067,004, **WALTER DECKER**, Zella St. BL. Saxe-Coburg, Germany. Solid frame, double action, hammerless revolver with square butt and trigger that travels horizontally backward (not pivoted) its lower portion being fixed to a rod that passes to sear in rear of frame. This allows trigger to be placed in front of cylinder and allows easier gripping and less recoil.

August 19, 1913, No. 1,070,582, **JOHN M. BROWNING**, Ogden, UT. The .45 Colt automatic, improved (butt rounded at rear).

September 7, 1913, No. 1,072,335, **LOUIS F. KUNZ**, Chicago, IL. Safety pivoted on top frame of revolver may be swung in front of cocked hammer, an arm projecting down at right angles stopping hammer's travel until lever is swung aside.

October 28, 1913, No. 1,077,135, **EDMUNDO GUERRERO**, Antigua, Guatemala. A swing out cylinder revolver with helical slots on surface of cylinder and a stud in frame (below cylinder) to engage these, stud being operated by link bar actuated by trigger and pivoted inside frame.

December 2, 1913, No. 1,080,364, **CHARLES A. NELSON**, Utica, NY. (Ass'd. to Savage Arms Co.) Thumb safety for Savage automatic. Other points.

December 30, 1913, No. 1,082,961, **CHARLES W. LANG**, Utica, NY. (Ass'd. to Savage Arms. Co.) Improvements on Savage auto safety.

December 30, 1913, No. 1,082,969, **CHARLES A. NELSON**, Utica, NY. Improvements on Savage Automatic.

February 3, 1914, No. 1,085,698, **CHARLES A. NELSON**, Utica, NY. Improvements to Savage semi-automatic.

December 1, 1914, No. 1,119,124, **FREDERICK SMITH**, Worcester, MA. (Ass'd. to Harrington and Richardson.) Trigger locking device for Harrington and Richardson Auto.

December 15, 1914, No. 1,121,437, **WILLIAM J. WHITING**, Handsworth, England. Improved type of safety which operates on removal of magazine (in automatic pistol).

December 29, 1914, **EDWARD S. POMEROY**, Springfield, MA. (Ass'd. to Smith and Wesson.) Safety block slips between hammer and frame and then remains until retracted by pull on trigger.

January 12, 1915, No. 1,124,312, **GIOVANNI MONGARDI**, Boston, MA. Cylinder swings to right in horizontal plane, being pivoted at forward end of frame. Cylinder pin release on right side of frame below hammer. Auto ejection.

January 26, 1915, No. 1,126,201, **GRANT HAMMOND**, Hartford, CT. (Ass'd. one-third to A.C. Washburne, Pittsfield, MA., and one-third to Frederick G. Crane, Dalton, MA.) Improved hammer safety for automatic pistols.

June 1, 1915, No. 1,141,285, **WILLIAM L. SWARTZ**, Utica, NY. (Ass'd. to Savage Arms Company.) Grip safety for Savage semi-automatic pistol.

June 15, 1915, No. 1,143,470, **WILLIAM J. WHITING**, Handsworth, England. Magazine controlled safety.

June 15, 1915, No. 1,143,471, **WILLIAM J. WHITING**, Handsworth, England. Combination magazine and grip safety.

July 27, 1915, No. 1,147,860, **HUGO HENGELHANPT** and **LEOPOLD MOLLER**, Philadelphia, PA. Revolving rifle with new latch and a hammer locking device.

August 17, 1915, No. 1,150,566, **LEONARD WOODS**, St. Louis, MO. A seven shot, .22 caliber revolver in a watch case (with crystal) firing through the stem.

November 2, 1915, No. 1,159,004, **RUDOLF FROMMER**, Budapest, Hungary Germany. Divided firing pin for automatic arms.

January 25, 1916, No. 1,169,121, **EDWARD W. BRYN**, New York, NY. A 75-shot rifle for high power cartridges, cylinder moving vertically as well as rotating and bringing all charges successively opposite barrel (which is placed about opposite center of cylinder) due to a spiral arrangement of the chambers.

March 21, 1916, No. 1,176,003, **CORNELIUS VANDERBILT**, Jr. New York, NY. Four shot pepperbox with fixed barrels and rotary firing pin. Hammerless. Grasped in hand and fired by squeezing upon a reciprocating plate. (Manufactured by Shattuck Arms Co.)

May 2, 1916, No. 1,181,417, **JOSEPH H. WESSON**, Springfield, MA. Ejecting mechanism for hand ejector revolver which will eject either rimmed or rimless cartridges.

August 8, 1916, No. 1,193,889, **JOSEPH E. GADBOIS**, Springfield, MA. Revolving rifle with four shot cylinder fed by an endless chain from a cartridge reservoir in the hollow stock. For small cartridges.

November 7, 1916, No. 1,208,848, **JOHN HOWARD CARL**, Gilroy, CA. Trigger, bolt and hammer locked by a single safety (thumb operated) on left side of frame.

February 13, 1917, No. 1,216,001, **REGINALD F. SEDGLEY**, Philadelphia, PA. Combined rebounding, main, and trigger spring for "Baby Hammerless" revolver.

May 22, 1917, No. 1,227,531, **PROSPERO DONADIO**, Brooklyn, NY. (Ass'd. to Fred Biffar, Chicago, IL.) Hammer locks in connection with hammer pin in double action, break open revolvers.

July 31, 1917, No. 1,234,961, **GEORGE H. TANSLEY**, Hartford, CT. (Ass'd. to Colts.) Magazine controlled safety.

August 14, 1917, No. 1,236,608, **REGINALD F. SEDGLEY**, Philadelphia, PA. Baby hammerless revolver with cylinder removable to right and ejection through pressure on cylinder pin.

July 9, 1918, No. 1,271,678, **LUCIUS N. DIEHM**, West Hartford, CT. Manually operated magazine pistol with magazine in stock. Exactly like the conventional hammerless automatic pistol.

November 26, 1918, No. 1,285,954, **LUCIUS N. DIEHM**, West Hartford, CT. (Ass'd. in part to D.A. Baldwin, B.C. Stone, Fred I. Hodge and Franklin Turner, of Middletown, CT.) Improved trigger mechanism for automatic pistols.

January 7, 1919, No. 1,290,855, **JOSEPH H. WESSON**, Springfield, MA. (Ass'd. to Smith and Wesson.) Automatic pistol with vertical handle containing magazine and semi-rotary bolt; hammer with cocking spur.

January 14, 1919, No. 1,291,688, **WALDO E. ROSEBUSH**, Appleton, WI. Automatic pistol constructed to take cartridges of different sizes (lengths) in same barrel, and of different calibers by interchange of barrels. Semi-automatic or automatic fire may be had. Hammer gun.

August 12, 1919, No. 1,313,073, **PROSPERO DONADIO**, Brooklyn, NY. Button (grip) tang safety for double action, break open revolvers.

August 26, 1919, No. 1,313,912, **EUGENE, G. REISING**, East Hartford, CT. (Ass'd. to Hartford machine Gun Co., Hartford, Ct.) Improved firing pin.

June 8, 1920, No. 1,342,923, **CARLOS POREIRA PINTO**, Buenos Aires, Argentina. Double action operation made more accurate by having a hard pull to cock arm and revolve cylinder and a light one to discharge it. (Shows foreign solid frame type.)

October 5, 1920, No. 1,355,067, **BERTIL A.B. RYDBERG**, Skofde, Sweden. Magazine controlled safety. Impossible to insert empty magazine.

October 12, 1920, No. 1,355,424, **JOHN D. PEDERSON**, Jackson, WY. Automatic pistol which somewhat resembles in outline the Remington Hammerless. Fixed barrel.

October 12, 1920, No. 1,355,425, **JOHN D. PEDERSON**, Jackson, WY. Similar arm of slightly different type.

August 9, 1921, No. 1,387,369, **JOSEF DUFFEK**, Steyr, Aust. Gun lock for automatic pistols.

September 6, 1921, No. 1,389,944, **GEORGE H. GARRISON**, Bridgeport, CT. Magazine controlled safety attached to trigger (for Remington automatic pistol.)

November 15, 1921, No. 1,396,969, **PETER MARINO**, Providence, RI. Superposed barrel rifle of two calibers with cylinder containing two rows of chambers, one set for each caliber, two hammers and two triggers.

November 22, 1921, No. 1,398,085, **BENJAMIN H. BALL**) (Ist. Lt., O.D.U.S.A.) Washington, DC. Revolver to be attached to muzzle of service rifle for use in trench warfare.

February 7, 1922, No. 1,405,765, **LUCIUS N. DIEHM**, West Hartford, CT. (Ass'd. one-half to B.C. Stone, Middletown, CT.) Hammerless single shot pistol on lines of an automatic. Looks like Fiala.

March 21, 1922, No. 1,410,269, **JOHN D. PEDERSON**, Jackson, WY. Grip safety for automatic pistols.

April 4, 1922, No. 1,411,800, **WILLIAM J. MOLLOY**, Chicago, IL. Replaces cylinder of a revolver with a blank made to fit flush with frame, with firing pin running through it, to be actuated by hammer of revolver. Replaces cylinder pin with an extractor to be pushed rearward manually. Blank is hinged at bottom and swings laterally to permit loading.

September 5, 1922, No. 1,427,967, **JOSEF NICKL**, Oberndorf, Germany. (Ass'd. to Waffenfabrik Mauser.) Recoil operated, automatic pistol with ring hammer and safety.

December 5, 1922, No. 1,437,889, **THEODORE H. EICKHOFF**, Cleveland, OH. (Ass'd. to Auto Ordnance Corp., N.Y.) Improved "Submachine gun."

December 26, 1922, No. 1,440,332, **F. CHARLES A. BRAUNGER**, Elizabeth, NJ. Manually actuated action for firearms. (Cut resembles auto.)

May 8, 1923, No. 1,454,454, **HENRY ROSIER**, Lowell, MA. Hammerless, open top, solid frame revolver with folding trigger and stocks folding into very compact recess.

July 10, 1923, No. 1,461,387, **JOHN WILLIAM FEARN**, Birmingham, England. Improved firing pin safety for Webley type auto pistol, located on left side of slide.

September 4, 1923, No. 1,466,749, **JOHN D. PEDERSON**, Jackson, WY. and **CRAWFORD C. LOOMIS**, Ilion, NY. (Loomis ass'd. to Remington Arms Company.) Magazine controlled safety for Remington automatic pistol.

November 20, 1923, No. 1,475,037, **EDWARD THIMGREN**, Popotla, Mexico. Thumb trigger for Luger type automatic pistol.

April 22, 1924, No. 1,491,635, **JOHN D. RUSS**, Spencer, W. VA. A recoil operated mechanism to keep cylinder and barrel in close contact at moment of discharge. Acts on crane arbor (in hand ejector type revolvers) through a spring controlled wedge.

August 5, 1924, No. 1,503,568, **GEORGE WALTHER**, Zella-Mchlis, Germany. Grip safety.

November 25, 1924, No. 1,516,835, **RUDOLF FROMMER**, Budapest, Hungary. Device to prevent automatic fire from pistol designed for semi automatic work.

December 2, 1924, No. 1,518,027, **ULRICH VOSMAK**, Antigo, WI. Front cylinder pin lock to be added to Colt type swing out cylinder revolvers in addition to usual rear release. Device located on right side of forward end of frame, below barrel.

December 30, 1924, No. 1,521,286, **ERNEST L. FISHER**, New Brunswick, NJ. Trigger device to counteract tendency to pull to right or left. Portion of trigger engaged by finder presents a rotary element to compensate for lateral drag.

May 12, 1925, No. 1,537,369, **ITACUBURI DEL ROSARIO**, Paraguay. Revolver with square, auto like stocks and hammer in form of reciprocating bolt, like bolt action rifles. Double action. Ejection mechanism not pictured.

INDEX OF INVENTORS

N.B. The figures in brackets indicate the number of entries for the inventor on that page.